Maps from the Age of Discovery
Columbus to Mercator
by
KENNETH NEBENZAHL

TIMES BOOKS

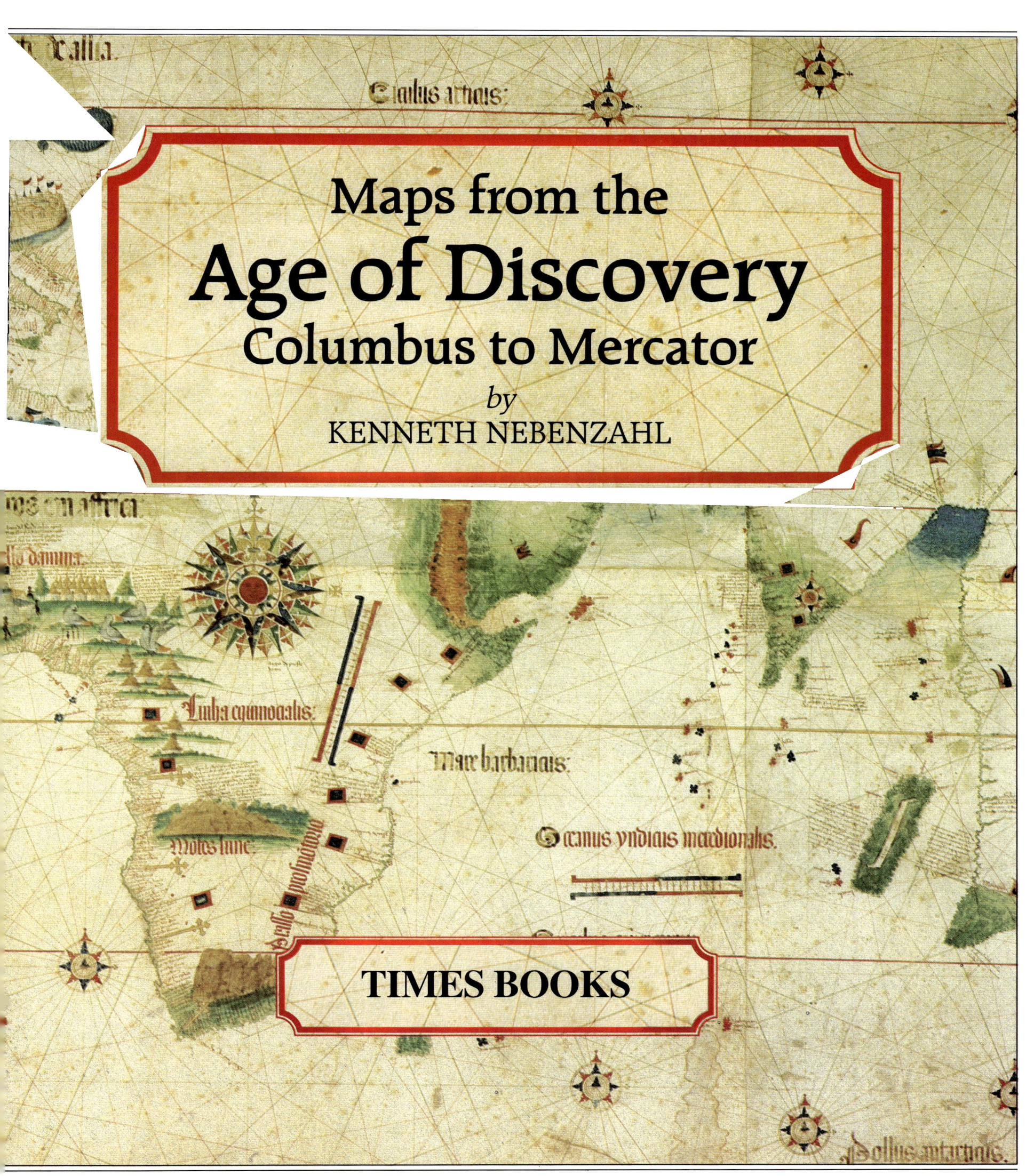

Maps from the Age of Discovery
Columbus to Mercator

by

KENNETH NEBENZAHL

TIMES BOOKS

ACKNOWLEDGMENTS

his book, even more than its predecessors, owes its existence to the tireless efforts and unselfish commitment of time by my wife, Jossy. From conception to completion, she worked through endless drafts and edited the material at all stages. The book simply would not have been written without her, and it is to Jossy that it is gratefully dedicated.

David Woodward of the University of Wisconsin–Madison kindly read the manuscript and made numerous suggestions for improvement. The advice and cooperation of many people, especially Laura C. Schmidt of Rand McNally, in acquiring the fine film necessary to produce images of the quality displayed here has been indispensable. It is a privilege to mention some of the others who helped most:

Robert W. Karrow, Jr., David Buisseret, Arthur Holzheimer, and photographer Kenneth E. Cain, the Newberry Library, Chicago; J. Brian Harley, University of Wisconsin–Milwaukee; Barbara McCorkle, Robert Babcock, and Patricia Middleton, the Bienecke Rare Book and Manuscript Library, Yale University; John Parker and Carol Urness, the James Ford Bell Collection, University of Minnesota; Roger E. Stoddard and Denison Beach, the Houghton Library, Harvard University; Susan Danforth, the John Carter Brown Library, Brown University, Providence; Sandra Sider, the Hispanic Society of America, New York; Beverly Carter, Paul Mellon Collection, Upperville, VA; William P. Frank, Henry E. Huntington Library, San Marino, CA; Tony Campbell and Malcolm Marjoram, the British Library, Map Library, London; Brian Thynne, National Maritime Museum, Greenwich; Monique Peletier, Bibliothèque Nationale, Paris; Robert Bassac, Service Historique l'Armèe de Terre, Paris; Hermann Maué, Germanisches National Museum, Nuremberg; Ernesto Milano, Biblioteca Estense, and Christine and Valerio Manfredi, Modena; Leonard E. Boyle, Biblioteca Apostolica Vaticana, the Vatican; Anna Lenzuni and Carla Guiducci Bonanni, Biblioteca Nazionale Centrale, Florence; Gian Albino Ravalli Modoni, Biblioteca Nazionale Marciana, Venice; E. Bos-Rietdijk, Maritiem Museum Prins Hendrik, Rotterdam; Urs Stocker, Dietikon-Zürich; Sabahattin Türkoğlu, Topkapi Saray Museum, Istanbul; Ildefonso Escribano, Museo Naval, Madrid; Robert Power, Nut Tree, California; Donald L. McGuirk; Paul Nebenzhal of New York; Linda Kreft, Paul Erling, and John J. Kennedy from my office; and the editorial staff of the World Atlas Division of Rand McNally.

Photograph credits

Jacket, Pages i, ii–iii, and 160: see Plate 11. Page vii: © 1979, The Metropolitan Museum of Art, Gift of J. Pierpont Morgan, 100. Page viii, top: from Thacher, John Boyd, 1904, *Christopher Columbus*, vol. 3, Cleveland: The Arthur H. Clark Co. Page viii, left: from Winsor, Justin, 1889, *History of America*, vol. 1, New York: Houghton, Mifflin and Co. Page viii, right: The Granger Collection, New York. Page 1: Thacher, 1904. Page 3, left: Courtesy of the James Ford Bell Library, University of Minnesota. Page 3, top: Winsor, 1889. Page 3, bottom: Germanisches National Museum. Plate 1: Biblioteca Apostolica Vaticana. Plate 2: Urs Graf Verlag. Plate 3: Courtesy of the James Ford Bell Library, University of Minnesota. Plate 4: Biblioteca Nazionale Marciana. Plate 5A: The Beinecke Rare Book and Manuscript Library, Yale University. Plate 5B: The British Library, Department of Western Manuscripts. Plate 6: Germanisches National Museum. Plate 7: Private collection. Plate 8: Bibliothèque Nationale. Page 26: Courtesy of the Newberry Library. Page 27, top: Courtesy of the James Ford Bell Library, University of Minnesota. Page 27, bottom: © Rand McNally. Plate 9: Courtesy of the Newberry Library. Plate 10: Museo Naval de Madrid. Plate 11: Biblioteca Estense. Plate 12: From Von Wieser, 1893. See Bibliography. Plate 13: Bibliothèque Nationale. Plate 14: The British Library, Map Library. Plate 15: Arthur Holzheimer collection. Plate 16: Facsimile courtesy of the Newberry Library. Plates 17A & 17B: National Maritime Museum. Plate 17C: Arthur Holzheimer collection. Plate 18: Courtesy of the John Carter Brown Library, Brown University. Plate 19: Courtesy of the Newberry Library. Plate 20: Topkapi Saray Museum. Plate 21: Ohio State University. Plate 22: Bibliothèque Nationale. Plate 23: Bibliothèque Nationale. Page 73, left: Private collection. Page 73, right: Private collection. Plate 24: Courtesy of the Newberry Library. Plate 25: By permission of the Houghton Library, Harvard University. Plates 26A, 26B, 26C, & 26D: The Beinecke Rare Book and Manuscript Library, Yale University. Plate 27: Courtesy of the Hispanic Society of America. Plate 28: Biblioteca Apostolica Vaticana. Plate 29: Biblioteca Apostolica Vaticana. Plate 30: Arthur Holzheimer collection. Plate 31: Courtesy of the James Ford Bell Library, University of Minnesota. Plate 32A, 32B, & 32C: Private collection. Plate 33: Private collection. Plate 34: Bibliothèque Nationale. Plate 35: By permission of the Huntington Library. Plate 36: The British Library, Department of Western Manuscripts. Plate 37: Ministère de la Défense, Service Historique de l'Armée de Terre/Photon S.A.—Studio Littré. Plate 38: The British Library, Department of Western Manuscripts. Plate 39: The British Library, Map Library. Page 125, top: Private collection. Page 125, bottom: Private collection. Plate 40 (detail): Maritiem Museum Prins Hendrik. Plate 40: Bibliothèque Nationale. Plate 41: Courtesy of the Newberry Library. Plate 42: Robert H. & Margaret C. Power. Plate 43: Courtesy of the Trustees of the British Museum. Plate 44: Courtesy of the Trustees of the British Museum. Plate 45: Paul Mellon collection. Plate 46: The National Maritime Museum. Plate 47: The National Maritime Museum. Plate 48: Paul Mellon collection. Plate 49A & 49B: Courtesy of the Newberry Library. Plate 50: Courtesy of the Newberry Library.

Maps from the Age of Discovery

Published in 1990 by Times Books Ltd.,
16 Golden Square
London W1R 4BN

Published in 1990 in the USA
by Rand McNally

General manager: Russell L. Voisin
Managing editor: Jon M. Leverenz
Editor: Elizabeth G. Fagan
Designer: Vito M. DePinto
Production and photo editor: Laura C. Schmidt
Production managers: John R. Potratz, Patricia Martin

All rights reserved. No part of this publication may be reproduced, stored in a retrieval system, or transmitted, in any form or by any means — electronic, mechanical, photocopied, recorded, or other — without the prior written permission of Rand McNally & Company.
Printed in Genoa, Italy.

ISBN 0 7230 0358 0

British Library Cataloguing in Publication data
Maps from the age of discovery: Columbus to Mercator.
1. Maps, history
912.09

TABLE OF CONTENTS

Introduction	vi

PART I:
The Cartographic Tradition Inherited by Columbus — 2

Claudius Ptolemy, World Map, Florence, 1474	4
Abraham Cresques, The Catalan Atlas, Majorca, ca. 1375	6
Zuane Pizzigano, Nautical Chart, Venice, 1424	9
Fra Mauro, World Map, Murano, 1459	12
Henricus Martellus Germanus, World Map, Florence, ca. 1489	15
Martin Behaim, Terrestrial Globe, Nuremburg, 1492	18
Donnus Nicolaus Germanus, Hispania, Ulm, 1482	20
"The Christopher Columbus Chart," Portolan Sea Chart, ca. 1492–1500	23

PART II:
Columbus and His Contemporaries Change the Map — 26

Map of the Discoveries of Columbus, Basel, 1493	28
Juan de la Cosa, World Chart, Santa Maria (Cádiz), 1500	30
The "Cantino" Planisphere, Lisbon, 1502	34
Bartolommeo Columbus & Alessandro Zorzi, Map of the Equatorial Belt, Italy, ca. 1503–1506/1516–1522	38
Nicolo Caveri, World Chart, Genoa, ca. 1504–05	40
Giovanni Matteo Contarini, World Map, Florence, 1506	44
Johannes Ruysch, World Map, Rome, 1507	50
Martin Waldseemüller, World Map, Strassburg, 1507	52
Francesco Rosselli, Marine Chart & World Map, Florence, ca. 1508	56
Vesconte de Maggiolo, World Map, Naples, 1511	60
Pietro Martyr d'Anghiera, Map of the Indies, Seville, 1511	60
Piri Re'is, Chart of the Ocean Sea, Gallipoli, 1513	62
Martin Waldseemüller, *Terre Nove*, Strassburg, 1513	64
Lopo Homem with Pedro Reinel, Northern Indian Ocean, Portugal, ca. 1519	66
Lopo Homem with Pedro Reinel, East Indies, Portugal, ca. 1519	67

PART III:
Filling in the Features of the Earth — 72

Hernando Cortes, Gulf of Mexico Map & Mexico City Plan, Nuremburg, 1524	76
Juan Vespucci, World Map, Italy, 1524	77
Antonio Pigafetta, Maps from Magellan's Great Voyage, Place of Origin Unknown, ca. 1525	80
Juan Vespucci, World Map, Seville, 1526	84
Gerolamo da Verrazzano, World Map, Place of Origin Unknown, 1529	88
Diego Ribero, World Map, Seville, 1529	92
Diego Ribero & Giovanni Battista Ramusio, The New World, Venice, 1534	96
Sebastian Münster, The New Islands, Basel, 1546	98
Battista Agnese, World Map, Venice, 1542	100
Battista Agnese, Oval World Map, Venice, 1542	102
Sebastian Cabot, World Map, Antwerp, 1544	104
The "Vallard" Chart, Dieppe, before 1547	108
Pierre Desceliers, World Map, Arques (Dieppe), 1550	112
Guillaume Le Testu, East Coast of North America, Florida, & the Greater Antilles, Le Havre, 1556	116
Diogo Homem, The North Atlantic, London, 1558	120
Abraham Ortelius, World Map, Antwerp, 1564	121

PART IV:
Europe's Colonial Era Begins — 124

Gerardus Mercator, World Map on Mercator's Projection, Duisburg, 1569	126
Georg Braun & Frans Hogenberg, Plan of Cuzco, Cologne, 1572	130
Jodocus Hondius, World Map, London, ca. 1589	132
John White, *La Virginea Pars* & *La Virgenia Pars*, London, 1585–1586	136
Baptista Boazio, *The Famouse West Indian Voyadge....*, Leiden, 1588	140
Baptista Boazio, *Hispaniola*, Leiden, 1588	144
Baptista Boazio, *Cartagena*, Leiden, 1588	144
Baptista Boazio, *St. Augustine*, Leiden, 1588	145
Cornelis de Jode, *Quiviriae Regnum* & *Americae Pars Borealis*, Antwerp, 1593	152
Edward Wright, World Chart on Mercator Projection, London, 1599	156

Bibliography	161
Index	165

INTRODUCTION

aps drawn from classical, medieval, and early Renaissance sources all affected the thinking of Christopher Columbus. He based his daring decision to sail westward to reach the Far East on geographical knowledge developed and passed down by scientists and philosophers for over fifteen centuries. These early "cosmographers" were unaware that an enormous portion of the world, now known as the Western Hemisphere, lay between the Old World and Asia. The European discovery of the West Indies, North, Central, and South America eventually was made by Columbus beginning in 1492, and the handful of explorers who were to follow.

In the late thirteenth century, Marco Polo had brought back tales of fabulous riches he had seen in the East. He recounted such wonders as the golden-domed temples and palaces in Cathay and Cipangu (China and Japan). These fabled lands boasted great quantities of magnificent precious stones, pearls, and silks fit for royalty. Tropical islands dotting the seas were covered with trees and bushes bearing spices. These tales sparked Europe's never-ending demand for Oriental luxuries and provided the incentive that launched the Age of Discovery.

The dream of a sea route to the Orient had not always preoccupied Western Europeans. In Roman times the two great civilizations of China and the West had been linked by the tortuous overland silk road. This caravan route wound 4,000 miles (6,500 kilometers) around desert regions and over high mountain passes from Xian to the Levant. Goods were then shipped across the Mediterranean to ports of the Roman Empire. Merchants traded silks for wool, gold, and silver to carry on the return journey.

With the decline of Rome, this trade faltered and the silk road was abandoned. Later the Islamic expansion into Asia blocked all trade with the East. In the late 1200s, the Mongols allowed traders such as the Polo family to resume travel along the old silk road. However, since the Ottoman Turks controlled strategic areas of western Asia and the eastern Mediterranean, the trade that developed was a mere trickle compared to Roman times.

Prince Henry the Navigator finally furnished a Portuguese solution to the trade route problem and provided the first impetus for European discoveries in the rest of the world. By 1420, Henry had founded his renowned school for mathematicians, astronomers, navigators, cartographers, and instrument makers at Sagres, the "sacred point" at Portugal's and Europe's most southwesterly corner.

Henry's dreams of exploration and discovery were to have a strong impact on the course of history. He planned to outflank the Islamic domination of African and Asian trade by establishing sea routes for his Portuguese caravels to reach the Orient. Although the Prince died before his plans were fully realized, the voyages he sponsored led to the discovery of the Azores, the Cape Verde Islands, and western Africa as far as the Guinea and Ivory coasts.

Columbus, who had gone to sea at age 20 in 1471 from his native Genoa, found himself shipwrecked five years later on the Portuguese coast, not far from Sagres. This fortuitous landfall has been referred to as "divine intervention" by those who feel Prince Henry's Portugal strongly molded Columbus's thinking and his eventual plans to sail westward to the Orient.

Columbus's cosmographical curiosity would have made him aware of the three kinds of maps known in his day. The first had its origins in the Greek world of the Roman era, as represented by Nicolaus's revival of Ptolemy's work in the 1470s (Plate 1). These maps exaggerated the Eurasian landmass and the length of the Mediterranean, minimizing both the circumference of the world and the size of the ocean between Europe and Asia. This distorted geogra-

No known portrait was made of Columbus during his lifetime. This oil painting attributed to Sebastiano del Piombo is dated 1519, just thirteen years after the great discoverer's death. It is considered the most reliable representation of Columbus.

phy encouraged Columbus to believe that it would take far less time to sail from Europe to Japan than the true distance actually required.

The second type of maps were medieval *mappaemundi* or world maps, widely known during the 1400s, that disseminated church dogma as well as geographical information. These maps were usually rectangular or circular, with Jerusalem at the center and an ocean surrounding the known world of Europe, Africa, and Asia. The culmination of this type was Fra Mauro's map, drawn in Venice in 1459 (Plate 4). This map combined traditional cartography with new information supplied by Portuguese and other explorers.

In the third category were the practical portolan sea charts used in Columbus's day by navigators, ship owners, and others involved in maritime affairs (Plates 3 and 8). These covered mainly the Mediterranean, occasionally extending beyond its shores. They were solely concerned with accurately mapping the coastlines. The most distinctive features of portolan charts were their great detail in depicting shorelines, the absence of information further inland, and the network of loxodromic or rhumb lines radiating across the seas from nautical compasses. These maps remain the true forerunners of modern sea charts.

In the final decade of the fifteenth century, European explorers, beginning with Columbus, dramatically changed those maps for all time. Sailing westward across the Ocean Sea, they encountered a huge landmass barring their way to the riches of the Orient. This barrier was what came to be known as America. The maps of that time indicated only a few islands between the western coast of Europe and the eastern shores of Asia. The navigators' problem in the Age of Discovery was how to find a way around or through this massive obstruction that appeared to stretch from Arctic to Antarctic.

For mapmakers, the great problem was how to accommodate these new discoveries by Columbus, Cabot, Vespucci, and their followers on conventional world maps. Columbus, first to find the West Indies and South America (which rightly should have been called Columbia), and

Top: *This 1575 woodcut by Tobias Stimmer, derived from a painting in the collection of Paulus Jovius, is known as the Jovian Portrait.*

Left: *Theodore de Bry's 1595 engraving of Columbus, believed to have been developed from the painting by Piombo on the previous page.*

Right: *A Spanish rendition of Columbus's landing on Hispaniola, originally published in Frankfurt by de Bry.*

John Cabot, who put North America on the map for his patron, Henry VII of England, believed they had reached an easterly appendage of the Asian continent. For them, no other explanation was possible. Their reports confounded cartographers for decades to come.

Almost a generation passed after Columbus's first voyage before mapmakers became confident that a previously unknown "fourth part of the world" existed in the separate continents of North and South America. This realization meant that the Ocean Sea was in fact two oceans—the Atlantic and the Pacific. As the discoveries continued, new maps emerged to fill in the blank spaces of the New World.

After Columbus's voyages, an ever-widening circle of explorers from Italy, Portugal, Spain, England, and France followed in rapid succession: Cabot in Newfoundland and Nova Scotia in 1497; Vasco da Gama in India the same year; Amerigo Vespucci on the South American coast; the Corte-Real brothers in Greenland and Labrador; and Balboa crossing the isthmus of Panama to see the Pacific in 1513, the year Ponce de León reached Florida. Then in 1519-22 Magellan's fleet provided the first real proof that the world was round—and the Pacific a vast expanse of ocean—when his ships circumnavigated the globe.

Spain and Portugal assumed command in the early decades of the Age of Discovery. With the aid of the Pope, they divided the overseas world between them. Beginning in 1493, a series of Papal bulls and treaties specified which nation governed which lands. England and France, offended at being excluded, did their best to establish overseas empires of their own.

Each nation produced maps to help in its overseas efforts, but Spain and Portugal considered their maps the secret property of their respective crowns. Revealing their contents to outsiders was a capital offense. The two Iberian nations maintained master manuscript planispheres in the government department that administered their overseas affairs. These royal charts were updated when seagoing captains and pilots returned from their expeditions. Occasionally, as in the case of Cantino and Ribero (Plates 11 and 29), such charts were secretly copied and smuggled out of the country. Diego Ribero compiled charts known as the Spanish *padrón real*; the beautiful example shown in this atlas survives in the Vatican Library.

The signature of Columbus from a letter written to his son, Diego, on December 3, 1504, upon returning from his fourth voyage.

The full range of these historic voyages of discovery is depicted in the four sections that follow. The atlas begins with Ptolemy's map and brings the cartographic history of the great discoveries to the end of the sixteenth century, when Europe's colonial era was well established. By then the coastlines of the world had been mapped with comparative accuracy. Except for the extremely remote regions of the Arctic, Antarctic, Alaska, Australia, and New Guinea, continental contours are recognizable (Plate 50). Most of the interiors of North and South America, Africa, and Asia would have to wait for later generations of explorers to map.

The growing colonial presence of Europeans in the New World is also reflected in maps of that time. Among the most impressive are conquistador Hernando Cortes's map of Mexico City; the plans of Cartagena, Santo Domingo, and St. Augustine from Sir Francis Drake's historic voyage; the original watercolor maps of Sir Walter Raleigh's Roanoke colony; and a contemporary plan of Cuzco, the Inca capital captured by Pizarro.

This atlas traces the progress of mapping the world during the age of the great discoveries by displaying superb reproductions of original works that tell the dramatic story. These plates have been produced, with few exceptions, from film made directly from the original manuscripts, rare woodcuts, and engravings. Beginning with maps such as those Columbus had seen, the book moves through the succeeding years as new information from each voyage of exploration and discovery found its way back to Europe. This unique collection enables the reader to view the world as it appeared to those who were living when the greatest changes in geographical history took place.

PART I

The Cartographic Tradition Inherited by Columbus

Christopher Columbus, while lacking formal schooling, managed to master the writings of many important ancient and medieval geographers. His son and biographer, Ferdinand, asserted that Columbus knew works by Aristotle, Seneca, Ptolemy, Strabo, and Pliny, which confirmed his belief that he could sail westward from Europe directly to "The Indies." Ferdinand's claims are supported by surviving books from Columbus's library that contain extensive marginal notes written in the Admiral's own hand.

Columbus also read Marco Polo's account of his travels in Asia and was familiar with the cosmographer Pierre d'Ailly's *Imago Mundi* (Image of the World). Polo had added a 30 degree eastward protrusion to the true coastline of Asia, thereby decreasing the supposed width of the ocean between Asia and Europe. Almost two thousand years earlier, Aristotle was reported to have written that the ocean between Spain and the Indies could be crossed in a relatively short time. Cardinal d'Ailly, drawing from Roger Bacon's theory, referred to this key statement in his *Imago Mundi*. These miscalculations, combined with Ptolemy's errors of longitude, encouraged Columbus to believe that a westward journey to the East was possible.

In the 1470s, Paolo Toscanelli, the Florentine physician and cosmographer, was the earliest known Renaissance supporter of a westward voyage. He contended that the Far East could be reached more directly by sailing west than by rounding the Cape of Good Hope and crossing the Indian Ocean. Toscanelli accepted Marco Polo's earlier claim of the elongated Asian continent. Columbus corresponded with Toscanelli, who sent an encouraging reply along with a copy of a letter and map he had prepared at the request of Afonso, King of Portugal, outlining his ideas. These documents deeply affected the course of Columbus's life and the history of the world. Although Toscanelli's letter has survived, his historic map was lost.

Part One reviews the three types of maps that formed Columbus's geographic heritage: the classical world of Claudius Ptolemy, the medieval world images of Fra Mauro and Abraham Cresques, and the traditional sailors' portolan charts of Zuane Pizzigano. This section also displays the 1489 Henricus Martellus "modernized" Ptolemaic world map. It includes Marco Polo's data and revises the coastline of Africa based on reports from Portuguese explorers. Martellus's map served as a source document for Martin Behaim, who summarized known geographical information on his famous globe in 1492.

These maps reveal the cartographic elements that encouraged what Columbus later called *La Empresa de las Indias*, the Enterprise of the Indies—a great adventure that forever changed the maps of the world. Part One also includes the 1482 woodcut map of the Iberian Peninsula. This map allows the reader to follow the main events in Columbus's life as he travelled through Portugal and Spain before his historic First Voyage and between the succeeding three voyages.

A sixteenth-century woodcut representing the mythology of the sea prevalent in Columbus's time.

The first page of the 1485 Latin edition of Marco Polo's Travels. *Columbus's own copy survives in Seville with marginal annotations in his hand.*

Martin Behaim's globe, representing the world as known to Europeans the year Columbus embarked on his first voyage of discovery.

CLAUDIUS PTOLEMY, WORLD MAP, FLORENCE, 1474

his world map epitomizes the sum of classical Greek geographical knowledge. Ptolemy lived in the second century A.D. in Alexandria, a Greek cultural center at the mouth of the Nile in Egypt. He left for posterity seminal works on astronomy, mathematics, and geography that were crucial to the cartography of the Great Discoveries.

After the fifth century A.D. Ptolemy's science, like so much of ancient Greek learning, was all but forgotten in Europe. Not until the Italian Renaissance did enthusiasm for classical knowledge revive. Just after 1400, a Byzantine manuscript of Ptolemy's geography was brought to Florence and translated into Latin. This version of Ptolemy's work was often copied and widely circulated, including magnificently illuminated examples such as the one illustrated.

The editor was Donnus Nicolaus Germanus, a German Benedictine working in Florence. Nicolaus has become so identified with the early editions of Ptolemy's atlas that it is difficult to tell where the classical science of Ptolemy ends and Nicolaus's renaissance interpretation begins. For example, the wind boys, personifying the principal wind directions, appear named and characterized as they did in the time of Aristotle.

Within the limits of the world known to ancient Europeans, the map is remarkably accurate. But as it proceeds beyond Britain and the Baltic in the north, the Canaries in the west, the upper Nile and Persian Gulf in the south and east, its accuracy diminishes. Subequatorial Africa and the Indian Ocean, completely landlocked, are characteristic inaccuracies. Nicolaus added a touch of modernism by appending to the outline of Ptolemy's world a gore showing an early fifteenth-century version of Scandinavia and the Arctic Ocean.

Although Ptolemy ably summed up the works of Eratosthenes, Hipparchus, and especially Marinus of Tyre, he considerably underestimated the circumference of the earth. Ptolemy's calculations provided an encouragement to Columbus that might have proved fatal but instead turned out to be extremely fortunate for the explorer. While Eratosthenes had calculated the earth's circumference centuries earlier with precision, Ptolemy figured the distance to be almost thirty percent shorter than had his predecessors.

This smaller sized earth, combined with an elongated Eurasian landmass and longer Mediterranean Sea, helped to convince Columbus that if he sailed westward he could reach the Far East. On his first voyage he arrived in the West Indies in the time he had judged it would take him to reach Japan. Had there been no "New World" dividing the Ocean Sea and blocking his way to the Far East, however, Columbus's voyage would have ended in disaster on the open sea.

Ptolemaic maps such as this acted as a matrix for the development of Renaissance world maps. These "modern" maps, while modifying and improving Ptolemy's work, were still based on his cartography. Throughout the great period of world exploration about to unfold, Europeans kept returning to Ptolemy for reinforcement and comparison with new concepts

and new delineations. Ptolemy's atlases went into dozens of printed editions beginning in 1477 and continuing well after 1700.

As new lands were discovered, new maps were added to Ptolemy's original canon of twenty-seven. The early Ptolemaic maps were reprinted in those atlases even when they contrasted sharply with the modern works prepared by the editor of each ensuing edition. In this way, as has been well stated, "Ptolemy the

PLATE 1
Claudius Ptolemy
(ca. A.D. 90–168)
World map,
thought to have been
originally compiled ca.
A.D. 150
Florence, 1474
Illuminated manuscript
on vellum, 17 x 23.5 in.
(425 x 588 mm.)
Location:
Biblioteca Apostolica
Vaticana, Vatican City

ancient remained as modern as the latest discovery."

It is not known who drafted the first map of the world at the dawn of civilization, but every mapmaker since has drawn upon the work of predecessors. Columbus's thinking was conditioned by his exposure to three kinds of maps: the classical Greek cosmographer's world map such as Ptolemy's; the medieval Christian rectangular and disk maps that allowed the "flat earth" interpretations; and the practical navigators' coastal plans, known as portolan sea charts. All had their effect, but none were more important than those inspired by the revival of Ptolemy.

References: Bagrow/Karrow 1990, no. 33/32; Bagrow/Skelton 1964, 34-37; Brown 1950, ch. 3; Brown (comp.) 1952, no. 3; Cortesão 1969-71, 1:92-113; Destombes 1964, no. 55:48; Goldstein 1965; Harley & Woodward 1987, 1:ch. 2; Parry 1963, 9 passim; Penrose 1955, 4 passim; Polaschek 1959; Vignaud 1902, 74 passim; Wroth 1944, no. 1.

ABRAHAM CRESQUES, THE CATALAN ATLAS, MAJORCA, CA. 1375

he large "Catalan" world map is the most remarkable surviving cartographic monument of the Middle Ages. Its name is derived from the school of cartography in Majorca that used the Catalan language. The area covered by the map extends from the Ocean Sea in the west, including the Canary Islands and Madeira, to the islands east of the Asian coast. The Baltic Sea and southern Scandinavia appear in the north, and the southward extent runs across Saharan Africa through the Indian Ocean to the East Indies. This map represents the first major step towards the Renaissance planispheres that appeared after 1500.

The Catalan Atlas has been attributed traditionally to the famous Jewish map and instrument maker of Majorca, Abraham Cresques. It was commissioned by Peter of Aragon, King of this Mediterranean court at Majorca. Like Sicily a few centuries earlier, Majorca was a meeting place for Arabs, Jews, and Christians and was ruled by a patron of geography. Despite a strict papal ban against commerce with the Islamic world and the continuing desire of the Aragon dynasty to organize another crusade, a vigorous trade with the East flourished under Christian sovereigns.

The King of Aragon, responding in 1381 to a request from King Charles V of France for a copy of his best world map, sent to Paris this map by "Cresques the Jew." There is a conflict in dates appearing in inventories of Charles V's library, which has raised some question about whether the map should be attributed to Cresques. No satisfactory alternative mapmaker has been proposed, however, and Cresques is still presumed by many experts to be the author.

A celebrated book illuminator and cartographer, Cresques compiled a number of portolan charts and gained the patronage of the royal house. For this elaborate project he is said to have been assisted by his young son, Jehuda, later the chief geographer for Prince Henry "The Navigator" of Portugal. The elder Cresques was perhaps the greatest early Jewish mapmaker. His ethnic background allowed him to straddle European and Muslim cultures. It also enabled him to draw on many works of Arabic science that were translated first

PLATE 2
Abraham Cresques (presumed author, dates unknown)
World map known as "The Catalan Atlas" Majorca, ca. 1375
Illuminated manuscript on vellum; 12 sheets mounted on panels, each ca. 10 x 25 in. (250 x 650 mm.)
Location:
Bibliothèque Nationale, Paris

into Hebrew before appearing in Latin.

The most dramatic innovation on Cresques' map is the appearance of the Far East, which is a graphic depiction of Marco Polo's *Travels,* an account that circulated in manuscript throughout Western Europe. The fascination of Medieval Europeans with the Polo family's epic journeys in the opulent Orient and tales they brought back about Japan, the East Indies, and India, became one of the strongest stimulants to European overseas expansion. Marco's descriptions of castles of the Khans in Cathay, silk costumes and decorations, golden saddles, precious stones and pearls, spices and other Oriental exotica excited the desires of Europeans and sparked the search for trade routes to the Far East.

In areas where the Polos had visited and reported, Cresques' map considerably improves the theoretical cartography of Ptolemy. The peninsular shape of India emerges clearly for the first time, presenting a truer picture than that found on maps available a century later. Central Asia and China reflect the actual experiences of explorers and travelers and include divisions of the Mongol empire and many lakes, rivers, and cities previously unknown in the West. Elsewhere we find either classic Ptolemaic outlines or biblically inspired places such as the Land of Gog and Magog. Nevertheless, no better delineation of Asia was available in Europe until the world maps of the sixteenth century.

Society has been well served by the preservation of documents such as this that greatly enhance our understanding of the past. The Catalan Atlas provides a picture of the world as it was known in the period before the impact of Columbus's discoveries. Its profuse decoration, figures of the Polo family and their caravan, richly costumed Oriental potentates, and even Jayme Ferrer's ship from Majorca searching for the River of Gold off the North African coast, give insight into the thinking of Western Europe. The time was becoming ripe for overseas adventure.

References: Grosjean 1978; Harley & Woodward 1987, 1:314 passim; Kimble 1938; Mollat du Jourdin 1984, no. 8; Nebenzahl 1986, 46-49; Nordenskiöld [1897], 25-44, 58-59; pls. 11-14; Rey Pastor & Camarero 1960, 56-60.

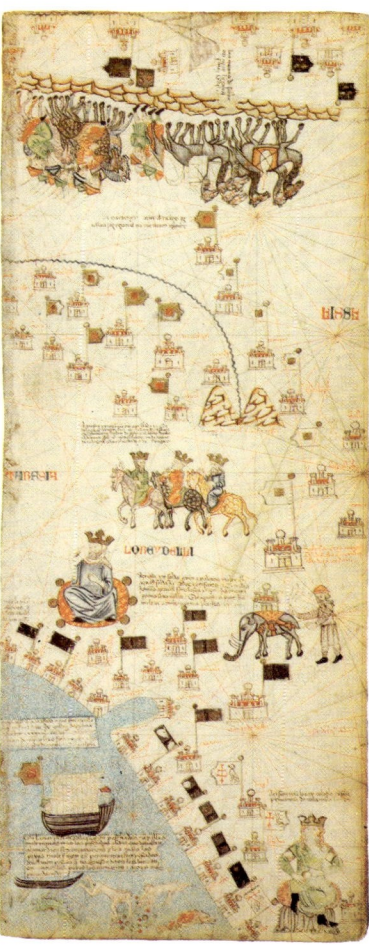

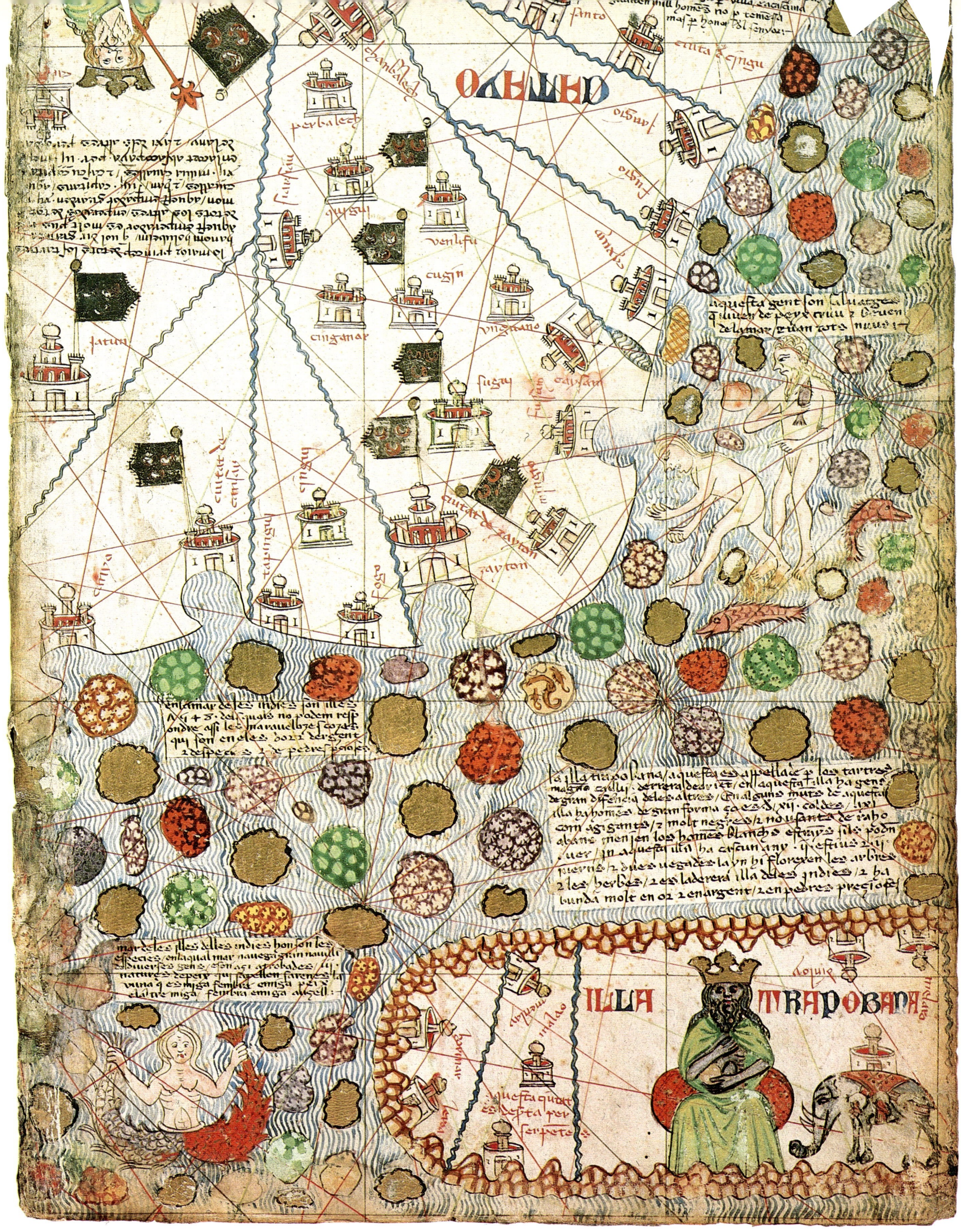

PLATE 2 (detail)
Abraham Cresques
Southeast Asia, reflecting the Marco Polo tradition

ZUANE PIZZIGANO, NAUTICAL CHART, VENICE, 1424

his is the earliest medieval map to concentrate on the Ocean Sea and the question of its mysterious islands. Here for the first time in the history of cartography a new group of islands appear, lands that exerted a strong influence on the course of history. Never before had there been a cartographical indication that sizable "countries" were to be found to the west of Europe. The largest, *Antilia*, is drawn as big as Portugal. It has been claimed that this is the first attempt to represent any part of what later became known as America.

Columbus's son Ferdinand mentions in the biography of his father that the Admiral had gathered information in Portugal, "particularly about that island called *Antilia* which lies 200 leagues westward of the Canaries and the Azores." Along with the opulence of the Indies reported by Marco Polo and other early travelers, the Atlantic islands became powerful magnets that compelled Columbus to seek a route to the Orient by sailing westward across the Ocean Sea.

These islands were not a new concept. Plato wrote about Atlantis in 400 B.C.; although his island may have been imaginary, it could have been based on factual knowledge. He may have learned of it from the Phoenicians, since that great sea-faring people had navigated as far as England and the Azores before 1000 B.C. Aristotle must have had some basis for his specific mention of calm seas filled with seaweed far out in the Atlantic, an obvious reference to the unique Sargasso Sea. Roman historians and geographers Seneca, Plutarch, Solinus, and Pomponius Mela continued to foster belief in lands west of the Mediterranean. There were ancient legends passed down for centuries about St. Brendan, the Irish monk who found a delightful island in the Atlantic in the sixth century, hundreds of years before the Norsemen went to Greenland and America.

An old and persistent Portuguese legend was of the "Seven Cities" (Antilia), founded in the eighth century by a bishop from Oporto. According to the story, when the Arabs invaded Iberia, he and other bishops fled to the "Island of the Seven Cities with gold amongst the sands and other riches," where they lived for years. These cities are represented by names appearing on *Antilia* in this delineation.

The Pizzigano portolan chart was discovered after World War II in the famous manuscript collection of Sir Thomas Phillipps, which was then being dispersed. There is no biographical information about the cartographer, who is presumed to be a descendant of the fourteenth-century Pizzigano family of mapmakers.

The chart shows the North Atlantic with coastlines of Ireland, England, Western Europe, the Iberian Peninsula, Majorca, and northwest Africa. Inscriptions follow in Venetian dialect, but names of the "new" Atlantic islands are in Portuguese. This suggests that the Portuguese named and presumably "discovered" the islands and that there could have been a Portuguese source for the Venetian chart.

Portolan sea charts, such as this, first appearing around 1300, represented a revolutionary break from classical and medieval geographical maps. They are distinguished by considerable detail along the known coastlines, stylized contours where direct knowledge was lacking, and an absence of information in the interior. A network of rhumb lines provided compass bearings for sailors to follow. The lines, also called loxodromes, give the direction for a ship to sail between two compass points. These charts were austere, practical, scientific aids, actually graphic representations of the written directions for seamen which had previously provided the sole guidelines for navigation.

Christopher Columbus rarely questioned speculative information if it tended to confirm his theory of sailing westward to reach "Cathay" (China). The chartmakers who followed Pizzigano continued the Atlantic island image. Columbus actually included a stop at "Antilia" in his itinerary on the way to Cipangu (Japan) and China.

References: Babcock 1922; Beazley 1897-1906; Bunbury [1883] 1959; Cortesão 1954; Kimble 1938; Parker 1955, 1-5.

PLATE 3 (detail)
Zuane Pizzigano
The island Antilia

PLATE 3
Zuane Pizzigano (dates unknown)
Nautical chart
Venice, 1424
Manuscript on vellum,
23 x 36 in. (570 x 900 mm.)
Location:
James Ford Bell Library,
University of Minnesota,
Minneapolis

FRA MAURO, WORLD MAP, MURANO, 1459

n the island of Murano in the Venetian lagoon at mid-fifteenth century lived the internationally renowned mapmaker Fra Mauro, who worked at the Camaldulian Monastery of San Michele. Although none of his early maps have survived, he had been a cartographer for close to twenty years when Alfonso V, King of Portugal, commissioned him to produce a large map of the world.

With the help of another well-known Venetian mapmaker, Andrea Bianco, the work was completed in April 1459. That same year the Venetian authorities requested a copy of the map. Because the King of Portugal's original has since perished, it is fortunate indeed that the Venetian copy of this, the great monument of early cartography, which immortalized Mauro's name, has survived.

It is the largest extant early world map—almost two meters in diameter. At first glance it seems to be simply a beautifully decorated example of a typical circular medieval *mappamundi*. A closer view reveals elements drawn from several traditions. The Mediterranean, Black Sea, and Iberian peninsula appear with modified and improved delineations. The contours of these areas are clearly derived from portolan sea charts. In particular the length of the Mediterranean has been shortened dramatically, resulting in a smaller European landmass. Although Ptolemaic touches are still present, and Mauro refers to Ptolemy, this map clearly has its roots in the fourteenth century.

Antecedents of Mauro's work were *mappaemundi* such as those by Vesconte and the even earlier Ebstorf and Hereford maps produced by medieval scholars. Probable Islamic influence is indicated by the map's orientation with south at the top, confusing to some because it seems the world is upside down. This approach originated in tenth-century maps of Arabic cartographers, possibly to display Mecca at the top of the world.

Most interesting is the new information depicted. The northern part of the west African coast is improved, perhaps in response to information sent by the Portuguese King to Mauro. Here at mid-fifteenth century is a map that shows Africa as a separate continent with the Ocean Sea joining the Indian Ocean. Although Vesconte in the previous century allows for such a possibility, his coastlines are so traditionalized and his pre-Marco Polo Asia so distorted as not to challenge the Ptolemaic concepts. Early Renaissance versions of Ptolemy showed Africa to only 10 or 15 degrees below the equator, after which it went off the bottom of the map. The east coast of these delineations was joined to a mysteriously extended southeastern peninsula of Asia, making the Indian Ocean an inland sea.

In 1459, over ten years before Portuguese navigators are thought to have reached the middle of the west African coast, Mauro's map showed it would be possible to sail around the southern tip of Africa to the East Indies. This idea was not confirmed until four decades later when Vasco da Gama first reached India (Calicut) during his voyage of 1497-98.

In addition to the influences of Ptolemy's geography, medieval *mappaemundi*, Islamic maps, and the portolan sea charts, Mauro's map displays extensive Asian information derived from Marco Polo's *Travels*. The Marco Polo geography had appeared on the Catalan atlas (Plate 2), but it is on Mauro's map that the greatest impact of Polo data is shown, prior to the arrival of the printing press.

Mauro's richly decorated manuscript features scores of towns and cities throughout the world, most symbolized pictorially by clusters of Renaissance Venetian edifices, together with portraits of potentates and depictions of exotic fauna. The imagination of the designer, having been restrained in the known world of southern and western Europe and the Mediterranean, is given full rein in the more remote regions of Africa, Asia, and the Arctic lands. The vision of wealth and splendor in the Far East that Columbus, Cabot, and Vasco da Gama had in mind and which motivated their historic voyages can be seen on Fra Mauro's production.

This great map is both the ultimate artifact of medieval cartography and a prime example of the transition to early modern mapping.

References: Bagrow/Skelton 1964, 72-3; Boorstin 1983, 154-55; Cortesão 1969-71, 2:172-81; Crone 1968, 53-61; Destombes 1964, no. 52:14; Harley & Woodward 1987, 1:315 passim; Wroth 1944, 108-09, no. 12.

PLATE 4
Fra Mauro (d. 1460)
World map
Murano (Venice), 1459

Illuminated manuscript on ox hide, 78 x 77 in. (1960 x 1930 mm.)

Location:
Biblioteca Nazionale Marciana, Venice

PLATE 4 (detail) Fra Mauro, India and Southeast Asia, oriented with south at the top

HENRICUS MARTELLUS GERMANUS, WORLD MAP, FLORENCE, CA. 1489

enricus Martellus was the one mapmaker who linked the early Renaissance cartography, just emerging from medieval constraints, to mapping that reflected the discovery of the New World. Little is known of this important German cartographer, probably from Nuremberg, who worked in Italy from 1480 to 1496, and produced a number of important manuscript maps.

Martellus's world delineation, drawn in Florence and circulated by an engraved version prepared by Francesco Rosselli, helped to change the world. It is believed that one copy of a Martellus map found its way to Nuremberg and inspired Martin Behaim to make his famous globe (Plate 6). Another copy may have reached Christopher Columbus in Spain. They depicted graphically the theory that Japan was but 3500 miles (5635 kilometers) westward, and only 1500 miles (2415 kilometers) further lay the shores of Cathay. Columbus thus had documentary support for his beliefs about oceanic distances from his readings of earlier cosmographers, Cardinal D'Ailly and Paolo Toscanelli. This encouraged him to press on with his plan to sail west to reach "the Indies."

The map was constructed on the projection of Claudius Ptolemy, the classical Greek scholar. Ptolemy's geographical writings, largely disregarded in the Christian Middle Ages, became the basis for Renaissance geography. The Martellus delineation included some Ptolemaic dogma in its continental contours but significantly modified and improved upon the ancient model. Martellus revised the Ptolemy world map based on Marco Polo's information on Asia, and he incorporated the recent Portuguese voyages to Africa. His is the earliest map to show the African continent as described by Bartholomew Dias, who rounded the Cape of Good Hope on his voyage of 1487–88.

The huge Martellus manuscript world map at Yale University (Plate 5A) is supremely important because it shows the division of latitude *and* longitude into degrees. This enables scholars to trace Columbus's thinking with some measure of certainty; he used this map to confirm his idea that Japan was only 90 degrees west of Lisbon, when it was actually more than twice that far.

The Yale example shows more of the Ocean Sea in the Far East than does the British Library manuscript. Among the thousands of islands Marco Polo reported off the coast of Asia, an enormous Sumatra and Java are found in the south, while to the northeast is the huge island of Cipangu (Japan). In the Indian Ocean, the islands of Madagascar and Zanzibar, rather poorly drawn, add an intriguing aspect. Some historians claim that their presence indicates the map cannot be dated before 1498, when Vasco da Gama returned with news of his voyage to India. Others, pointing out how crudely these islands are portrayed, assert that since this represents more Marco Polo information there is no conflict with the map's accepted date.

Martellus's shape of the world represents the most complete knowledge of the day. The map is remarkable for its exciting new information, although being imperfect because of its acceptance of classical and medieval antecedents. It was the most accurate delineation available to Martin Behaim when he constructed his globe exhibiting the pre-Columbian world. Columbus himself could find no better map to show him the way to Asia.

References: Bagrow/Skelton 1964, 106-07, 224; Campbell 1987, 72-74, 213, 217; Cortesão 1969-71, 1:294; Crone 1961; Davies 1977; Destombes 1964, 229-34, pls. 37-38; Harley & Woodward 1987, 1:187 passim; Kish 1966; McCorkle 1985, no. 1; Shirley 1983, no. 17; Vietor 1962; Wroth 1944, no. 18.

PLATE 5A
Henricus Martellus Germanus (fl. 1480–96) World map, including Japan at upper right Florence, ca. 1489 Illuminated manuscript on paper, 48 x 72 in. (1200 x 1800 mm.)
Location:
The Beinecke Rare Book and Manuscript Library, Yale University, New Haven

PLATE 5B
Henricus Martellus Germanus
(fl. 1480–96)

World map
Florence, ca. 1489

Illuminated manuscript on paper,
11.8 x 18.5 in. (300 x 470 mm.)

Location:
The British Library, London

MARTIN BEHAIM, TERRESTRIAL GLOBE, NUREMBERG, 1492

ehaim is an enigmatic character whose name and impact on history may be larger than deserved. He exaggerated his exploits and associations, and early chauvinistic biographers duly recorded his legendary, Baron Münchhausen–like career.

The cartographer was born into a prosperous upper-class merchant family that two centuries earlier had emigrated from Bohemia (hence the name Behaim). Most of Martin Behaim's life was spent as a merchant and commercial agent stationed at various European cities, including Lisbon. In Portugal he married the daughter of the Flemish governor of the islands of Fayal and Pico in the Azores.

Ravenstein's definitive monograph on Martin Behaim bristles with frustration at the incomplete and contradictory information revealed by his thorough research. Many of Behaim's assertions of greatness cannot be confirmed. Most of these are set out in legends on the globe itself. They include Behaim's relationship with the King and court of Portugal and also with the Holy Roman Emperor. Behaim's exploits as commander of one of the two ships on Cão's expedition to the west coast of Africa, which he claims discovered the Cape of Good Hope before Bartolemao Diaz, are disputed, as is his Portuguese knighthood. Finally, his status as a leading astronomer, mathematician, and cosmographer, disciple of his fellow Nuremberger, Johann Müller (Regiomontanus), has aroused some skepticism.

Regardless of these uncertainties, Behaim did design and supervise the construction of this twenty-inch globe that bears his name. The cartographic delineation is essentially based on the world map of Henricus Martellus (Plate 5). It presents the world as Christopher Columbus viewed it when he set off on his first voyage. Behaim represents the cosmographical ideas underlying Columbus's urge to sail westward to reach

PLATE 6
Martin Behaim (1459–1507)
Terrestrial globe
Nuremberg (Germany), 1492
Illuminated manuscript,
vellum gores on a plaster sphere,
mounted to metal meridian
and horizon rings, in a tripod stand
Diameter, 20 in. (507 mm.)
Location:
Germanisches National Museum,
Nuremberg

the Indies. The rich decoration and beautiful colors are not what make this globe historic. Behaim uniquely preserves the ancient theories of the relationship of continents to oceans, passed down from Greek times; elaborated upon and perfected by medieval scholars such as Roger Bacon, Cardinal d'Ailly, and Paolo Toscanelli; and eventually adopted by Columbus.

The fabled island of Cipangu (Japan) is the target. The legend reads, "The most noble and richest island of the east, full of spices and precious stones. Its compass is 1200 miles... In this island are found gold and shrubs yielding spices." As shown here, the Ocean Sea from the Pillars of Hercules (the Strait of Gibraltar) to Cipangu is not a long journey. Due to errors of the cosmographers Columbus followed, and his interpretation of their calculations, he believed the distance between these points was the same as it actually is from Gibraltar to the Bahamas.

Columbus's landfall occurred in a "new world" he never expected, recognized, nor acknowledged. Its relative lack of portable material riches disappointed him, because he had been seeking Japan and Cathay. But his courageous voyage brought the Admiral glory beyond his imagination and personal ambition, neither of which was small. He proved that if one cannot be precisely accurate, having good fortune can change potential disaster to success and enduring fame.

Behaim's globe, the oldest surviving European terrestrial sphere, allows the most comprehensive view of the world as seen by Columbus.

References: Campbell 1987, 72, 213; Cortesão 1969-71, 2:205-06, 233-35; Mollat du Jourdin 1984, no. 20; Ravenstein 1908; Stevenson 1921, 1:47-52, 2:251; Wroth 1944, no. 17.

PLATE 6 (detail), *Martin Behaim, From eastern Atlantic Ocean to Indian Ocean*

DONNUS NICOLAUS GERMANUS, HISPANIA, ULM, 1482

The Iberian Peninsula was the theater in which the European side of the Christopher Columbus saga took place. His travels and travails in Portugal and Spain may be traced on this contemporary map. The *Hispania* was printed in 1482 as part of a "modern" series added to a German Renaissance edition of the Greek classical geography of Claudius Ptolemy. The 1482 atlas used the text and maps of Donnus Nicolaus, a German geographer working in Florence.

For his modern map of Spain, Nicolaus employed a manuscript by Henricus Martellus, another German–Florentine (Plate 5). Although without scale, grid, latitude, or longitude indication, it is a considerable improvement over the classical delineation. Dotted lines indicate borders of the kingdoms; and a rich nomenclature is recorded for towns, rivers, mountains, and capes.

Columbus was born at the Italian seaport of Genoa and went to sea in 1465 at age fourteen. Seven years later he joined the French pirate René d'Anjou, roving the Mediterranean from Gibraltar to the Greek islands. In 1476, while sailing under Guillaume de Casenove-Coullon in a naval battle against Genoa off the Portuguese coast, his ship caught fire. Columbus survived by swimming ashore supported by one of the ship's oars. The fact that he landed at the site of Prince Henry the Navigator's school for seamanship at Sagres is regarded as a sign of divine inspiration for his discovery voyages.

Columbus came under the spell of the cosmographical speculation rife in Lisbon, western-most port of the known world. He accompanied Portuguese voyages northward, possibly as far as Iceland, and sailed down the West African coast. While in Lisbon he read Marco Polo's *Travels* and medieval cosmographers' writings on the earth as a sphere. He became obsessed with the idea that by sailing westward he could reach the riches of Japan, China, and the East Indies more directly than by taking the arduous journey around the Cape of Good Hope and across the Indian Ocean.

The map identifies ports and cities important to Columbus's career. At left is Cape St. Vincent where he was delivered out of the sea into Portugal. Lisbon appears to the north in the Tagus estuary, home base for Columbus until 1484. In that year he presented his daring plan to the King of Portugal, who declined to sponsor the voyage.

Columbus then left Portugal and sailed for Spain, landing at the little port of Palos. He had his first audience with Ferdinand and Isabella in Córdoba, northeast of Seville. For four years Columbus was to follow the royal court from city to city: Salamanca, in west central Castile; Santa Fe, near Granada; eventually back to Palos, where three ships were outfitted for him to sail the Ocean Sea.

Upon his return from the New World in 1493, he was summoned to Barcelona and given a large fleet of seventeen ships with over one thousand men for his second voyage, which sailed from Cádiz in September

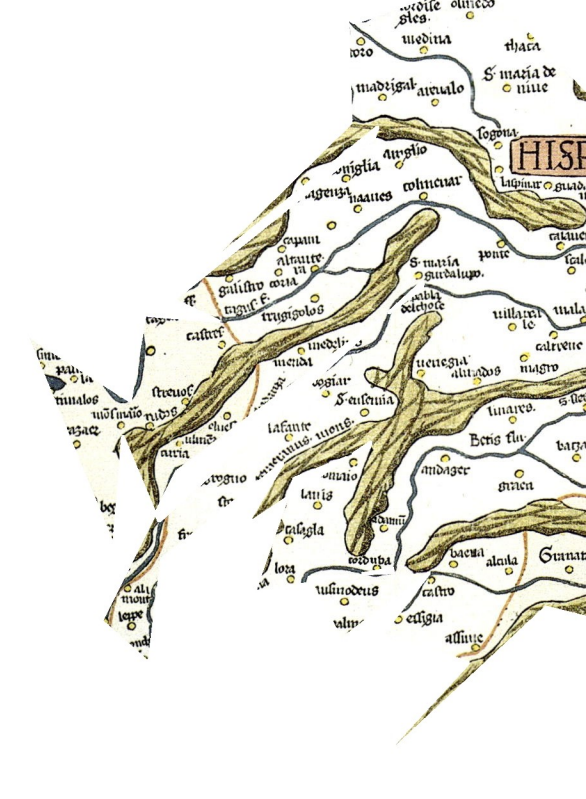

1493. The expedition was plagued by misfortune and by quarrels among the explorers. Returning over two years later, Columbus hurried to the King and Queen at Burgos to defend himself against reports that he was unable to govern his Hispaniola colony. The third expedition departed from Sanlucar, downstream from Seville. It was even more disastrous than the second, and ended with Columbus's return to Cádiz in November 1500 in chains.

His final voyage, planned in Seville, was harrowing and exhausting. Although he made important

PLATE 7
*Donnus Nicolaus Germanus
(ca. 1470–90)
Hispania*
Ulm (Germany), 1482
Woodcut, colored by hand,
14.5 x 19.8 in. (368 x 503 mm.)
Location:
Private collection

explorations, particularly on the coasts of Honduras and Panama, Columbus could restore neither his fortunes nor his prestige. Upon his return to Sanlucar in November 1504, his physical and mental health broken, Columbus had to be transported by mule the sixty miles to Seville. The following May, after Isabella's death, he was received by Ferdinand at Segovia, north of Madrid, a considerable journey for the aging Columbus. The King was sympathetic but would not restore him as governor of the Indies.

The Admiral of the Ocean Sea, one of the world's greatest navigators and worst colonial administrators, died a few days later in Valladolid, far from Andalucia, long his Spanish home province. All four historic voyages were conceived, organized, provisioned, launched, and ultimately concluded within the triangle comprising Palos, Seville, and Cádiz.

References: Almagià 1948, 27-31; Bagrow/Karrow 1990, no. 33:k.1; Brown (comp.) 1952, no. 37; Campbell 1987, no. 208; Skelton 1963, v-xi.

PLATE 8 (detail)
"The Christopher Columbus Chart"
Mappamundi *and cosmographical diagram*

"THE CHRISTOPHER COLUMBUS CHART," PORTOLAN SEA CHART, CA. 1492–1500

In 1924, Charles de La Roncière, the renowned French historian of exploration and cartography, attributed to Christopher Columbus a portolan sea chart that has been discussed and debated ever since. Although Columbus was an accomplished mapmaker, scholars have been frustrated in their attempts to confirm who actually created this unsigned document, originally acquired by the French national library in the nineteenth century.

The sea chart displays a classic delineation of the greater Mediterranean area, supplemented by the Atlantic coast stretching from southern Scandinavia to the mouth of the Congo River (named Rio Poderoso by Diogo Cão in 1484). It has particularly rich nomenclature down the African coast, where Columbus is thought to have made at least one voyage with the Portuguese. To the east, the Black Sea and Red Sea are included. Westward is a series of islands—some real, some imaginary—from the Arctic to the Gulf of Guinea. Below the compass rose in the North Atlantic lie three islands, "Isles of the Seven Cities." This was the Portuguese name for the islands that other Europeans called "Antilia."

In the neck, or narrow portion, of the parchment, a small circular world map centering on Jerusalem and surrounded by celestial rings symbolizes the geocentric concept of the universe, commonly accepted at that time. It is most unusual to have a practical navigator's chart juxtaposed with a cosmographical plan. One of the accompanying lengthy notes in Latin announces that the world map, or *mappamundi*, although drawn on a plane, should be considered spherical. Displaying the earth in this manner underscores the transitional character of the map from medieval to Renaissance thinking.

This circular *mappamundi* is also noteworthy for showing southern and eastern Africa more accurately than does either the Martellus map or the Behaim globe. It implies that information is included from Vasco da Gama, the Portuguese navigator who discovered India, even though he did not return to Europe until 1499. The treatment of the west, south, and east coasts of Africa suggests the map's Portuguese origin. Latin is used for the numerous lengthy annotations. The 250 place-names, however, appear in their Portuguese form, and many Portuguese-controlled areas display the Lusitanian flag. A reference at the Cape Verde Islands mentions their discovery by the Genoese, an intriguing fact considering Columbus's birthplace.

The surviving remains of Columbus's library include his revelatory marginal notes, particularly in his copy of Cardinal d'Ailly's cosmography, *Imago Mundi*. Monique de La Roncière, having researched these notes, recently pointed out that Columbus referred to his "four charts on paper, all of which also contain a sphere." She also noted an error in the inscription on this chart, next to the Red Sea, which is identical to an error in one of Columbus's marginal notes.

The Spanish flag flying over Granada implies the map was completed after January 1492, when Spain captured that city from the Moors. There was no attempt to show the new discoveries reported from 1493 onward, as recorded on the Juan de La Cosa planisphere of 1500 and those which followed. This fact suggests a date for this map no later than the early 1490s.

It appears questionable that a chart with this degree of professional finish and decoration, heightened with gold and including elaborate vignettes of selected major European cities, was actually executed by Christopher Columbus or his brother, Bartolommeo. The style and emphasis do not appear to support the assertion. Certainly it could have been commissioned by Columbus, and his Portuguese contacts could have provided the new information, which had to have originated in Lisbon. The chart, with its *mappamundi* inset, remains a remarkable document of the discovery period. Although the attribution to the Admiral by the French scholars has merit, it has never been confirmed.

References: Cortesão 1969-71, 2:220; Destombes 1964, no. 51:26; Fite & Freeman 1926, no. 3; Mollat du Jourdin 1984, no. 21.

PLATE 8
"The Christopher Columbus Chart"
Place of origin unknown, ca. 1492–1500
Illuminated manuscript on vellum,
28 x 44 in. (700 x 1100 mm.)
Location:
Bibliothèque Nationale, Paris

PART II

Columbus and His Contemporaries Change the Map

Christopher Columbus made landfall in 1492 in the Bahamas, certain that the small, flat island of sand and coral belonged to a group of islands off the China coast. With the long, dangerous voyage behind them, the jubilant crews of Columbus's three ships spent two days at the island called *Guanahani* by the "Indians." Their joy quickly turned to disappointment. Not only did they find poverty widespread among the Indians but there were no fabulous cities with golden-domed temples and palaces, as Marco Polo had reported.

Columbus ordered his captains to sail on and locate "nearby" Japan. Their frustrating search for the riches of Asia continued throughout the rest of the First Voyage. They returned to Spain in March of the following year, somewhat battered and confused, eight months after embarking.

Historians during the past two centuries have debated vehemently over exactly which island in the Bahamas or Turks and Caicos is the true Guanahani, or San Salvador, as Columbus ceremoniously named it. Although Watling's Island has had the most support, Grand Turk, Samana Cay, Cat Island, Mayaguana, and Caicos all have been proposed as the original landfall site. The argument has been raging since the 300th anniversary commemoration in 1793. It is unlikely that a unanimous choice will ever be reached. Problems of translation and interpretation of existing documents, plus the absence of definitive data concerning winds and currents, will probably keep one theory from prevailing over any others.

Europe first learned of Columbus's historic voyage when his report to Ferdinand and Isabella was printed as a pamphlet and distributed. In the 1493 Basle edition, the publisher included a woodcut illustration of what was purported to be Columbus's ship. It was accompanied by a picture-map of the West Indies.

Manuscript delineation of the northwest coast of Hispaniola, attributed by historians of cartography to Christopher Columbus. If drawn by the Admiral, this is the earliest surviving map of the New World.

These illustrations, which open Part Two of the atlas, are the first images of the New World shown in Europe. They have become well-known icons of that epoch.

Columbus's return created a crisis for cartographers. How and where could his new discoveries be placed on world maps? The Admiral remained convinced that on his First Voyage he had sailed among the outer islands of Asia. On his forthcoming Second Voyage, he vowed to locate and visit Cipango (Japan) and Cathay (China).

The maps featured in Part Two reveal the complexity of incorporating bewildering fragments of a New World on the existing, pre-Columbian delineations. Spanish, Portuguese, and German maps and charts surviving from the period enable the reader, five hundred years later, to follow the mapmakers' attempts at solving these problems.

Columbus's First Voyage overturned Old World concepts of the earth's geography. Yet it would require several generations of cartographers to assimilate the astonishing information brought back from subsequent voyages by Columbus, his contemporaries, and his successors.

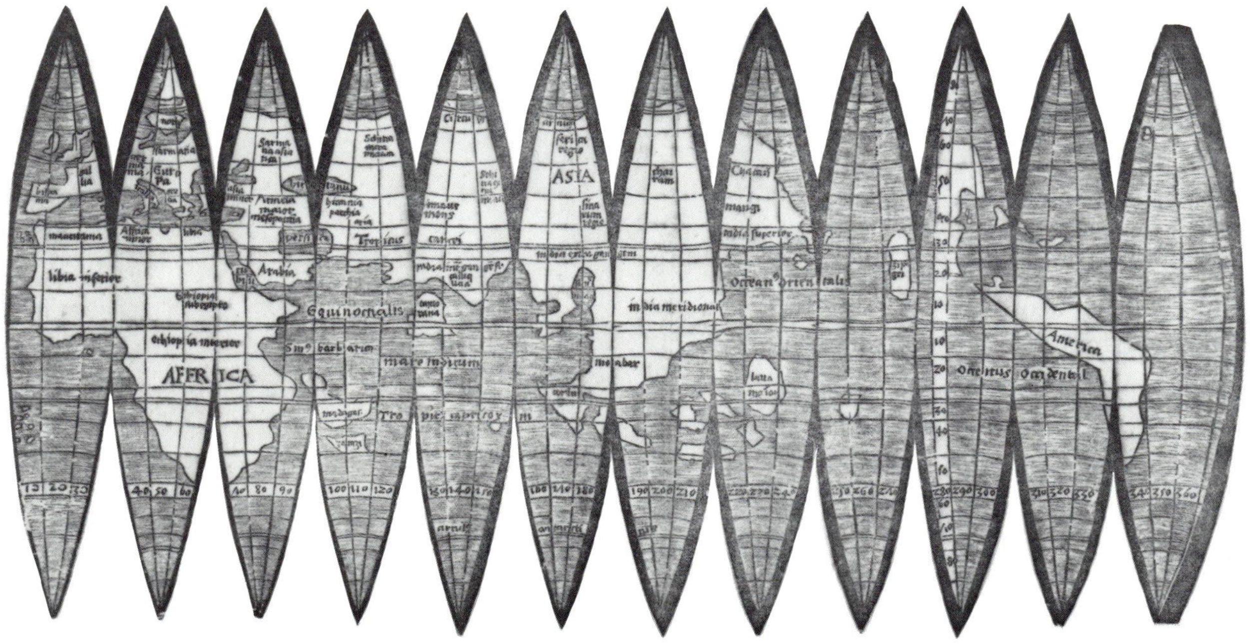

Gores for a terrestrial globe to accompany Martin Waldseemüller's Introduction to Cosmography, *1507. The earliest surviving map with the name "America."*

The Caribbean region on a modern map.

MAP OF THE DISCOVERIES OF COLUMBUS, BASEL, 1493

The most momentous of all discoveries, Columbus's encounter with the New World, was announced to Europe via a pamphlet. This eight-page leaflet became a best seller and informed a wide European audience for the first time of lands and peoples previously unknown. The impact of Columbus's discoveries upon Europe was intensified by the printing press, established only a few decades earlier. Even though the Norse had gone to America, and perhaps later Portuguese or other Europeans might also have crossed the ocean, most of Europe remained unaware of those earlier efforts.

The letter Columbus sent to his patrons, Ferdinand and Isabella, bearing the news of his return from the Ocean Sea and the discoveries of the "Islands of the Indies beyond the Ganges," was composed more as a public announcement than as personal correspondence. This first page of European-American history was written in Spanish and printed in Barcelona in March or April 1493. The first printing, which survives only in the New York Public Library collections, was hastily prepared from a poor copy of the manuscript. Immediately afterwards it was improved upon by a Latin translation from a better and more accurate copy of the original report.

Altogether, Columbus's *Letter* appeared in nine Latin editions during the year following the initial printing in Spanish. These were issued in Rome, Paris, Antwerp, and Basel. The text was translated into Tuscan verse by a Florentine poet, and five editions of the poem were published in 1493. By 1497 there was a German translation for sale in Strassburg. In this manner, some ten thousand copies of the account of Columbus's voyage were available within five years to the comparatively few literate Europeans.

The Basel editions were augmented by a series of woodcuts. These famous illustrations were totally imaginary but no less influential as the first pictures of the New World. Columbus's ship in one of the woodcuts appears as a multi-oared Mediterranean galley, and simple Indian dwellings were depicted as medieval European castles. These odd depictions can be understood if one realizes that this was the first wave of news about lands outside the long-familiar Euro-Asian arena.

The Admiral believed he had discovered the outer islands of Asia, and with perseverance, he could penetrate them to reach the Asian mainland. Although he did not reveal it in the *Letter,* his elation at reaching the new lands was tempered by disappointment; he had not found great riches nor located Cathay and the Grand Khan himself.

This picture-map is of primary importance as the first illustration of Columbus's discoveries. It shows his ship among the islands of the West Indies named "Fernanda, Hyspania, Ysabella, Salvatorie, and Conceptions Marie." It is a stylized view of the islands, that, with the caravel and its lone mariner standing on the poop deck setting his mainsail, is the first item of the 500-year continuum of European–American iconography.

References: Brown (comp.) 1952, 51; Campbell 1987, no. 70; Church 1907, no. 8; Harrisse 1866, no. 15; Hough 1980, no. 7; JCB 1919-31, 1:18; Sabin 1868-1936, no. 98923; Skelton 1958, no. 31; Tooley, Bricker & Crone 1968, 194.

PLATE 9A
Map of the discoveries of Columbus
Basel (Switzerland), 1493
Woodcut, 4.5 x 3.25 in. (115 x 80 mm.)
Location:
The Newberry Library, Chicago

PLATE 9B
Columbus at Hispaniola
Basel (Switzerland), 1493
Woodcut, 4.4 x 3 in. (113 x 75 mm.)
Location:
The Newberry Library, Chicago

JUAN DE LA COSA, WORLD CHART, SANTA MARIA (CADIZ), 1500

This large oxhide with its ink-and-water color map is a cartographic treasure rescued from obscurity in 1832 in a Paris antique shop. It is the first world map to include any of Columbus's discoveries and was drawn from personal experience by a participant, Juan de La Cosa. The chartmaker sailed as pilot with Christopher Columbus and later made three expeditions to the north coast of South America with Ojeda and Vespucci. He was killed in a skirmish with Indians in 1509 during an attempt to found a colony on the Spanish Main.

Columbus believed Cuba to be a peninsula extending from the mainland of Asia. In 1494 during his second voyage, he explored westward along Cuba's south coast. The journey was long and difficult. The ships' supplies ran low, the caravels badly needed repairs, and the crew were exhausted and ill. In an atmosphere of mutiny, Columbus finally gave in to the demands of his men and agreed to return to Hispaniola.

Before doing so, however, as insurance against blame for returning prematurely, he insisted that all officers and men, including La Cosa, sign before witnesses a most unusual document. It proclaimed that Cuba was indeed a peninsula and that proceeding further was a waste of time. The abrupt change of plans could have been caused by this crisis with the men and condition of the ships and supplies, or perhaps by Columbus's desperation to keep alive his geographical theory. It has never been certain whether he actually knew the truth about Cuba, but had he continued fifty miles further west, he would have discovered, much to his enormous disappointment and frustration, that Cuba was an island.

The chart omits latitude and longitude, and a different scale is employed for the New World than for the Old. America appears dramatically oversized, with the fragmentary information available to La Cosa imperfectly organized. This is contrasted with a smaller scale but accurate delineation of Europe, the Mediterranean, and West Africa.

On the chart, Cuba, Hispaniola, Jamaica, and the Lesser Antilles display numerous place-names; the distinctive fish-hook shape of Cuba became the hallmark of La Cosa's delineation. The chart includes detail along South America's northeast coastlines that reflects La Cosa's later adventures. In addition to his experience in the Caribbean, La Cosa drew upon a lost John Cabot map for parts of North America. In the vicinity of Newfoundland and Cape Breton is an east-west coastline with place-names which at that date could have come only from John Cabot's New World exploration. The names are derived from a map drawn by Cabot, which was known to have existed since it was mentioned in a letter sent from England to King Ferdinand of Spain. R.A. Skelton wrote: "The only map which unambiguously illustrates John Cabot's voyage of 1497... is the world map signed by Juan de La Cosa and dated 1500."

Although the delineation is crude, La Cosa was sophisticated enough to truncate Asia, thereby avoiding the question of whether Columbus and Cabot had reached the Far East or a "new world." Also, the illustration of St. Christopher covers the unknown region where the search for a passage to Cathay was soon to begin. At the narrow end of the document beneath St. Christopher's portrait, Juan de La Cosa's signature and the date 1500 appear.

Most scholars agree the map was drawn in 1500 and that this surviving chart is an early copy of the lost original; experts date the copy between 1502 and 1510. This world chart is the earliest cartographic record of the voyages of Christopher Columbus and John Cabot and possibly the most important surviving artifact of the Age of Discovery.

References: Campbell 1981, no. 3; Fite & Freeman 1926, no. 4; Ganong 1964, 8-43, 469-73; Harrisse 1892, 412-15; Hoffman 1961, 87-97; Kohl 1869, 151-55; Layng (comp.) 1956, 5. Morison 1942, 1:186-88 passim; Nordenskiöld [1897], 149; Nunn 1946; Skelton 1965, 16; Williamson 1962, 59ff; Winship 1900, no. 84; Wroth 1944, no. 19; 1970, 1:40-43.

PLATE 10 (detail), *Juan De La Cosa, The New World*

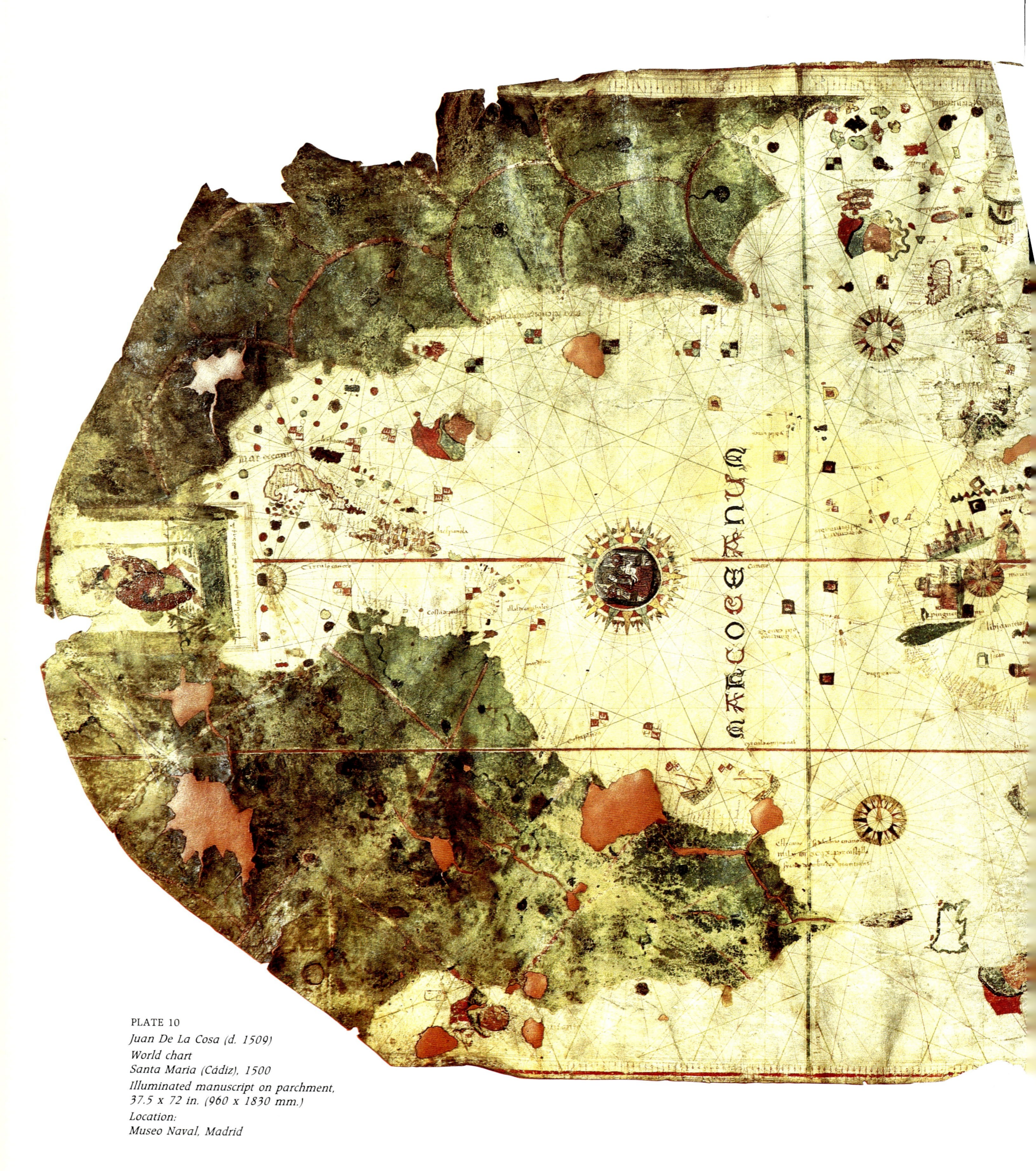

PLATE 10
Juan De La Cosa (d. 1509)
World chart
Santa María (Cádiz), 1500
Illuminated manuscript on parchment,
37.5 x 72 in. (960 x 1830 mm.)
Location:
Museo Naval, Madrid

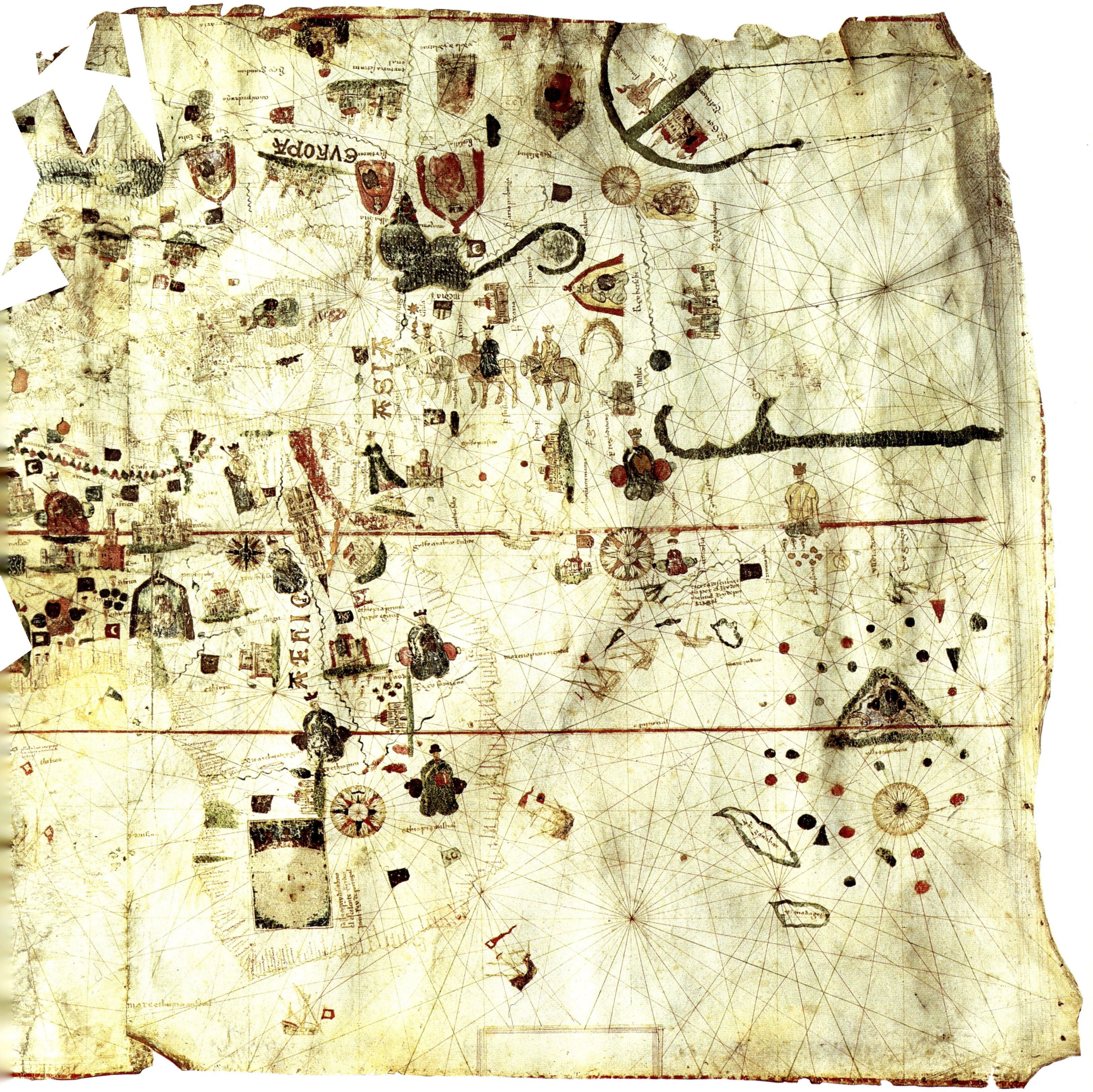

THE "CANTINO" PLANISPHERE, LISBON, 1502

As the Juan de La Cosa map graphically dramatizes the impact of Columbus on Renaissance Europe, the Cantino planisphere glorifies the achievements of the great Portuguese navigators of the same period including Vasco da Gama, Cabral, and the Corte-Real brothers. This, the first sea chart of the era of European trans-Atlantic discovery that can be precisely dated, is a manuscript born of controversy and intrigue.

The name of the cartographer remains unknown for interesting reasons. In the political atmosphere of this period, the need for anonymity was imperative. Success in the bitter rivalry between Spain and Portugal required that the new geographical data generated by discoveries in the East and West Indies be kept secret. Information from returning mariners was assembled by cartographers to form official charts for kings and their advisors. To copy or divulge the contents of these royal maps was a capital crime.

Nevertheless, there were leaks. An inscription in Latin on the reverse side of this map relates that "this sea chart of the islands recently discovered in the regions of the Indies has been presented to the Duke of Ferrara, Ercole d'Este, by Alberto Cantino." A Lisbon-based diplomatic agent of the powerful Italian Este family, Cantino secretly obtained this important document and smuggled it out of Portugal. It is said that Cantino interviewed Amerigo Vespucci, who had just returned from the New World, and acquired from him the new place-names that appear on the map.

Evidence of when an early manuscript was produced is essential in determining its priority and significance. For example, several key but undated maps were made in the first years of the sixteenth century, but the primacy of the 1502 Cantino world map is established on the basis of its date. The latest information here is a legend expressing fear that Gaspar Corte-Real, the Portuguese explorer, has perished in the North Atlantic. Because two of Corte-Real's ships returned to Lisbon and brought this news in October 1501, the map could not have been completed before then. Additionally, there is evidence that the Duke of Este received the map in November 1502. While the maker unfortunately remains unknown, the date is certain.

This earliest surviving Portuguese map of new discoveries in the East and West represents the known world at the exciting moment when Europe was learning of its actual extent. The American coastline remains fragmentary, because probes had been made only to the West Indies, Nova Scotia-Newfoundland, part of the South American coast, and possibly Florida. The stretches visited were recorded, but gaps in the coastline were not filled in until further explorations.

On Greenland a flag and inscription announce the Portuguese arrival. Farther west, majestic trees on the large island in the North Atlantic recall the description of the east coast of Newfoundland given by the Corte-Reals when they returned to Lisbon in October 1501. In the West Indies (*Antilhas*), the statement appears that "these are the West Indies of the King of Castille, discovered by Columbus...Admiral of these islands...at the command of the most high and mighty King don Fernando...."

Northwest of "Isabella" is an area, incomplete and partially off the map, that is perhaps the greatest unsolved cartographic puzzle of the period. Although "Isabella" strongly resembles Cuba, and the peninsula to the northwest could be Florida, there are several theories to the contrary. One is that the anonymous Portuguese mapmaker confused Spanish reports of the configuration of the newly discovered islands and duplicated Cuba; first as the island but also as the incompletely explored area to the northwest. Another interpretation considers "Isabella" to be Cuba but regards the peninsula as the Asian mainland Columbus and Cabot believed they had reached.

If we are seeing Cuba and Florida, no one knows from whom this information came, as Florida was not formally discovered until 1513. There is speculation that an early Amerigo Vespucci voyage may have been the source, or that an unknown Portuguese pilot could have unofficially sailed through Spanish waters before 1500 and coasted Florida. Lawrence Wroth and other historians of New World discoveries believe that both Cuba and Florida were depicted, and that Cantino's chart was the prototype for the important maps of the Lusitano-Germanic series. These delineations, such as the Waldseemüller wall map (Plate 16), did much to illuminate the New World for Europeans.

The Brazilian coast displays Portuguese flags and announcements of the landing in April 1500 by the Portuguese, Pedro Alvares Cabral. There is no reference to the arrival on the north part of the coast in 1499 of Vicente Pinzón, Columbus's early partner. Prominently shown is the line of demarcation of the Treaty of Tordesillas, signed by Spain and Portugal in June 1494. This is the oldest surviving map to bear this historic line. A meridian was drawn some 960 nautical miles west of the Cape Verde Islands that divided the entire world in two for the purpose of European overseas expansion. Spain was given the portion west of this line in the Atlantic and Portugal the east. Consequently, although Spain claimed most of America, the Portuguese controlled the East Indies and Brazil.

This extraordinary manuscript documents the epochal accomplishments of Portuguese mariners early in the Age of Discovery.

References: Cortesão 1969-71, 1:7-13; Harrisse 1892, 77 & 422; Hoffman 1961, ch. 7; Layng (comp.) 1956, 6; Lowery 1912, no. 3; Nordenskiöld [1897], 149-50; Stevenson 1903, no. 1; Wroth 1944, no. 21; 1970, no. 2.

PLATE 11 (detail), *The "Cantino" planisphere, The New World*

Oceanus occidentalis
Terra del Rey de portuguall
Has antilhas del Rey de castella:
Este he o marco dantre castella e portuguall
Toda esta terra he descoberta p mãdado del Rey de castella
Linha equinocialis:
Tropicus capricorni.
Pollus antarticus:
Mare oceanus:
Os montes claros em affrica:
Castello damina:

PLATE 11
The "Cantino" planisphere
Lisbon, 1502

Illuminated manuscript
on 3 vellum leaves, joined,
40 x 86 in. (1020 x 2180 mm.)

Location:
Biblioteca Estense,
Modena (Italy)

BARTOLOMMEO COLUMBUS & ALESSANDRO ZORZI, MAP OF THE EQUATORIAL BELT, ITALY, CA. 1503–06/1516–22

Christopher Columbus was marooned in Jamaica for almost one year during his fourth voyage. From there on July 7, 1503, he wrote a letter to King Ferdinand, reporting on his exploration of Nicaragua and Panama. A copy of the letter was brought to Rome in 1506 by Columbus's brother, Bartolommeo, who had accompanied the Admiral on this final voyage. Bartolommeo was seeking the Pope's support to persuade the King of Spain to grant a commission for colonizing and Christianizing the Central American coast.

Alessandro Zorzi, a Venetian who gathered accounts of explorers and travelers, was in Rome when Bartolommeo arrived. Zorzi, assisted by Bartolommeo, embellished an Italian translation of Columbus's letter with these three sketch maps. They appear as marginal illustrations and together comprise an equatorial zone map of the world. These surviving examples were included in a geographical manuscript written by Zorzi about 1522. The source of the delineation was a 1503 Christopher Columbus chart of Central America. That chart, now lost, had been with Bartholomew in 1506 and was reportedly seen by the contemporary historian, Peter Martyr, in 1516.

The cartographical concepts are complicated and represent a retrogression precipitated by the Columbus brothers' disappointment that no passage to Asia could be found in Central America. Before the fourth voyage, they presumed the New World was in fact a separate enormous island. Afterwards, however, they reverted to the belief that the mainland they explored was part of the Asian continent's eastern coast. The sketch maps clearly prove this point. Places Columbus visited along the Honduran coast on the fourth voyage are recorded on the first section as if they were on an Asian coastline west of the West Indies but attached to South America. The names on this coast are *Cariai, Carambaru, Bastimentos, Retrete,* and *Belporto.* On the second portion, these same names appear again along an indisputably Asian coastline.

The distance between Europe and Asia is grossly underestimated, a tenet basic to Columbus's thinking. The landmass in the northwest of the first section has a configuration similar to that in the "Cantino" map (Plate 11). This area has been interpreted variously as representing either Columbus's concept of Cuba, the peninsula of Florida, or perhaps the Asian mainland. Westward on the same continent, Columbus inscribed names for China from ancient maps by Ptolemy: *Serica,*

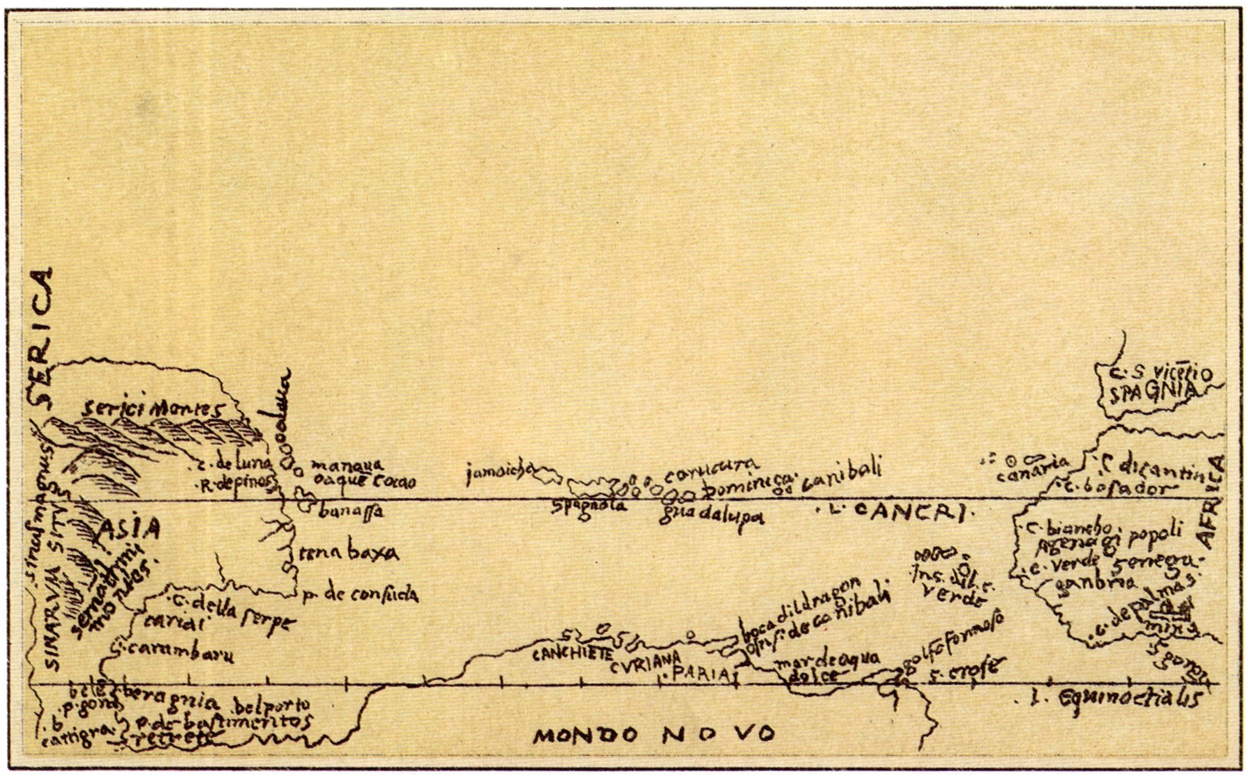

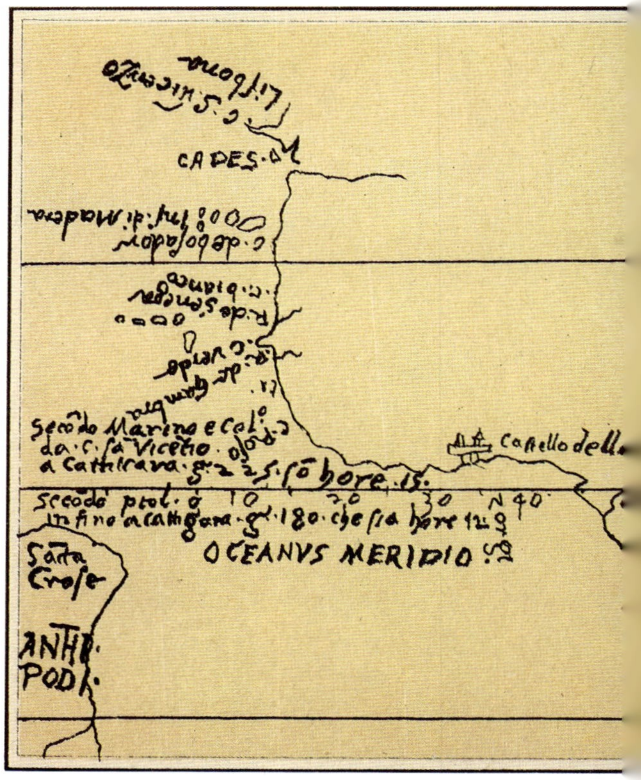

PLATE 12
*Bartolommeo Columbus (1460–1514)
& Alessandro Zorzi (dates unknown)
Sketch maps of the equatorial belt of the world
Italy, ca. 1503–06/1516–22
Pen and ink on paper, 3 sheets,
each 4 x 6.5 in. (100 x 165 mm.)
Location:
Biblioteca Nazionale Centrale, Florence*

Serici Montes, and *Sinarum Situs.* To the south Columbus imagined a narrow isthmus (Panama, later to be discovered by Balboa). West of the isthmus, he labeled the sea, *Sinus Magnus,* the classical name for the waters east of Asia. He even included a strait through the isthmus to account for the sea route Marco Polo had used to return from China two centuries earlier.

South America is called by Columbus *Mondo Novo* (The New World), a term generally credited to Vespucci. The shape of the coastline reflects the explorations of Columbus in 1498 and Ojeda in 1499. Among the few recorded place-names, "the Sea of Fresh Water" designates the mouth of the Orinoco.

The confusion experienced by the most well-informed minds in the early years of discovery is underscored on these sketches. The pre–Columbian world, represented by the Martellus map and the Behaim globe (Plates 5 and 6), was being forced to accommodate a "fourth part" of that world, America. The three "maplets" are treasured artifacts revealing the beliefs of Christopher and Bartolommeo Columbus.

References: Almagià 1948, 27-31; Bigelow 1935; Fite & Freeman 1926, no. 5; Layng (comp.) 1956, no. 15; Lowery 1912, no. 4; Nordenskiöld [1897], 167-69; Nunn 1924; 1952; Skelton 1958, no. 33; Von Wieser 1893.

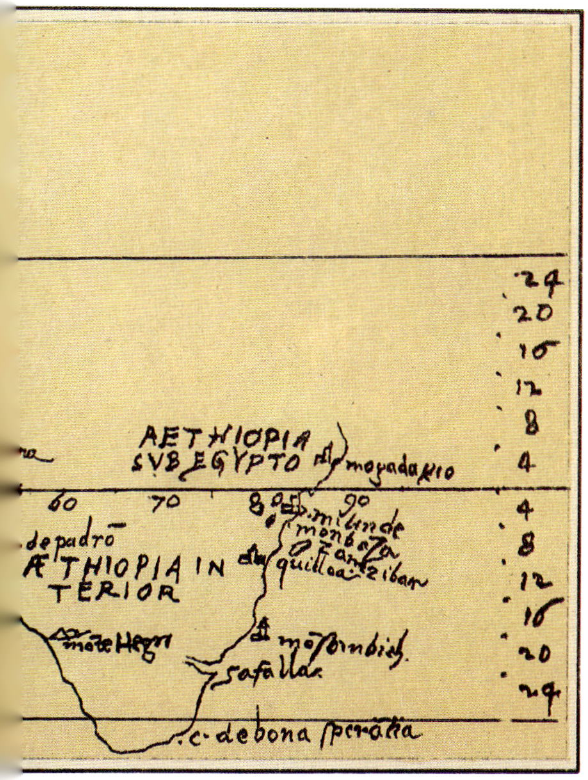

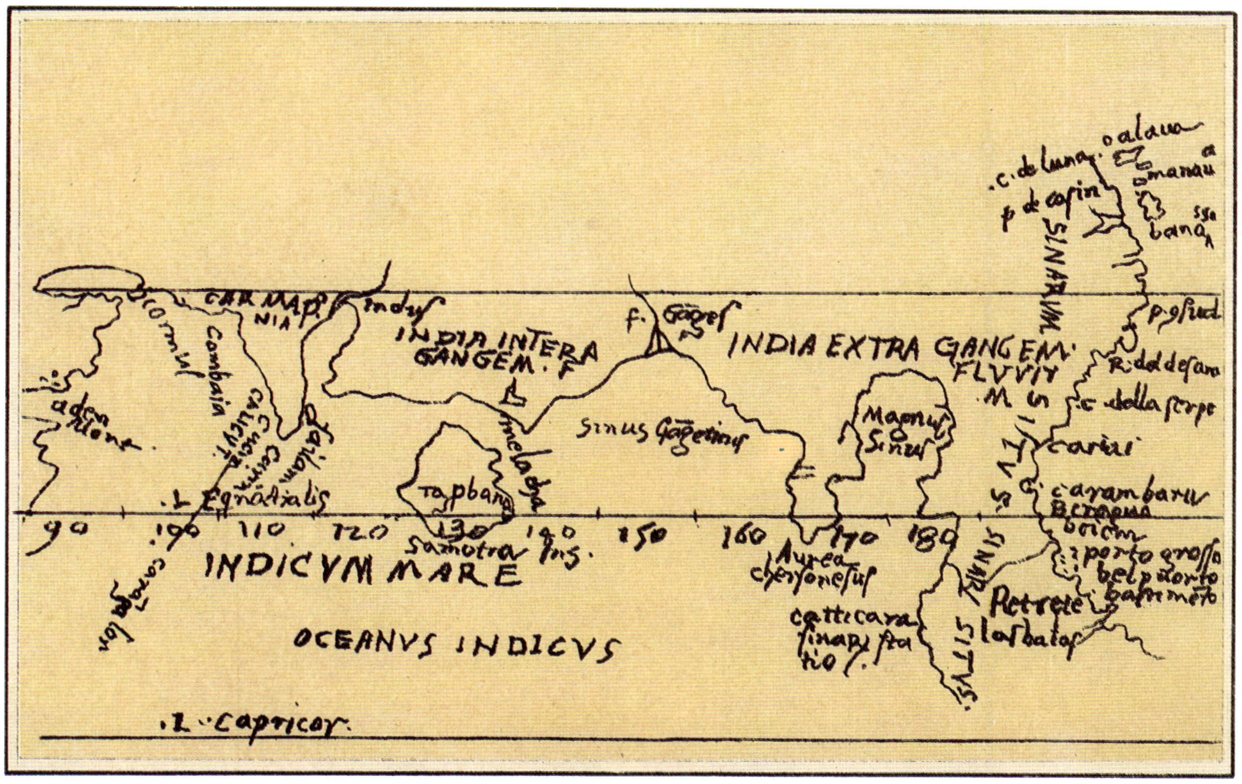

NICOLO CAVERI, WORLD CHART, GENOA, CA. 1504–05

The sea-faring city of Genoa, Columbus's birthplace, returns to center stage with this major cartographic contribution. The Genoese mapmaker, Nicolo Caveri, referred to by historians as Canerio until recently when his signature on this map was reinterpreted, has a place in history with this production. The Portuguese toponymy employed throughout shows that Caveri had access to Lusitanic prototypes. Its strong likeness to the "Cantino" planisphere (Plate 11) indicates that Caveri either used the "Cantino" chart or the two maps had very similar sources.

This great planisphere bridges the medieval and Renaissance worlds, as dramatized by the circular *mappamundi* at its center. Radiating out from the nucleus are rhumb lines that connect with a circle of compass roses. Rhumbs are also projected from these, forming a network covering the chart. Outside this circle six further points of intersection with compass roses inscribe yet another, though incomplete, concentric circle. This mesh of loxodromic lines has its origin in the late thirteenth century with the oldest of surviving sea charts.

If the *mappamundi* in the middle is a reference to the past, the latitude scale in the left margin is an innovation of great significance for future mapmaking. Since voyages across the seas had become a possibility, determining latitude accurately became essential to record compass directions. With this information navigators could return directly to home port and revisit newly discovered lands. Dead reckoning and a few latitude readings may have brought Columbus to America initially, but for regular travel more scientific methods were required.

Caveri and the "Cantino" map show similar contours for most parts of the world, although Caveri's mapping of the Red Sea is less accurate. In South America, particularly on the Brazilian coast, Caveri includes new place-names reported by two Portuguese expeditions. The first, in 1501–02, included Amerigo Vespucci; the second, commanded by Fernando de Noronha in 1503–04, attempted to establish a trading post in Rio de Janeiro Bay. Since Caveri had information from these voyages, the map probably dates from 1505. In North America he presents a new delineation of the Gulf of Mexico, with the peninsulas representing Yucatan and Florida. Although the relationships of Cuba, Yucatan, and Florida are only partly correct, Caveri's concept of the Gulf region was widely used for the next twenty years.

While the great Spanish planisphere by Juan de La Cosa (Plate 10) fell into virtual oblivion until its modern rediscovery, Caveri's chart was responsible for a continuous series of derivatives over the next twenty-five years, principally the 1507 twelve-sheet printed world map of Waldseemüller (Plate 16). These maps served to present the image of the New World to Europeans until news of further explorations of Ayllon, Verrazzano, and Gomez corrected and helped to complete the cartography of North America.

References: Giraldi (ed.) 1954-55, no. 13; Harrisse 1892, 77, 422; Hoffman 1961, ch. 7; Layng (comp.) 1956, no. 7; Lowery 1912, no. 2; Mollat du Jourdin 1984, no. 26; Nordenskiöld [1897], 150; Stevenson 1908; Wroth 1944, no. 20; 1970, no. 3.

PLATE 13 (detail), *Nicolo Caveri, The West Indies*

PLATE 13
Nicolo Caveri (dates unknown)
World chart, Genoa, ca. 1504–05

Illuminated manuscript on 10 vellum leaves, joined, 46 x 90 in. (1150 x 2250 mm.)
Location:
Bibliothèque Nationale, Paris

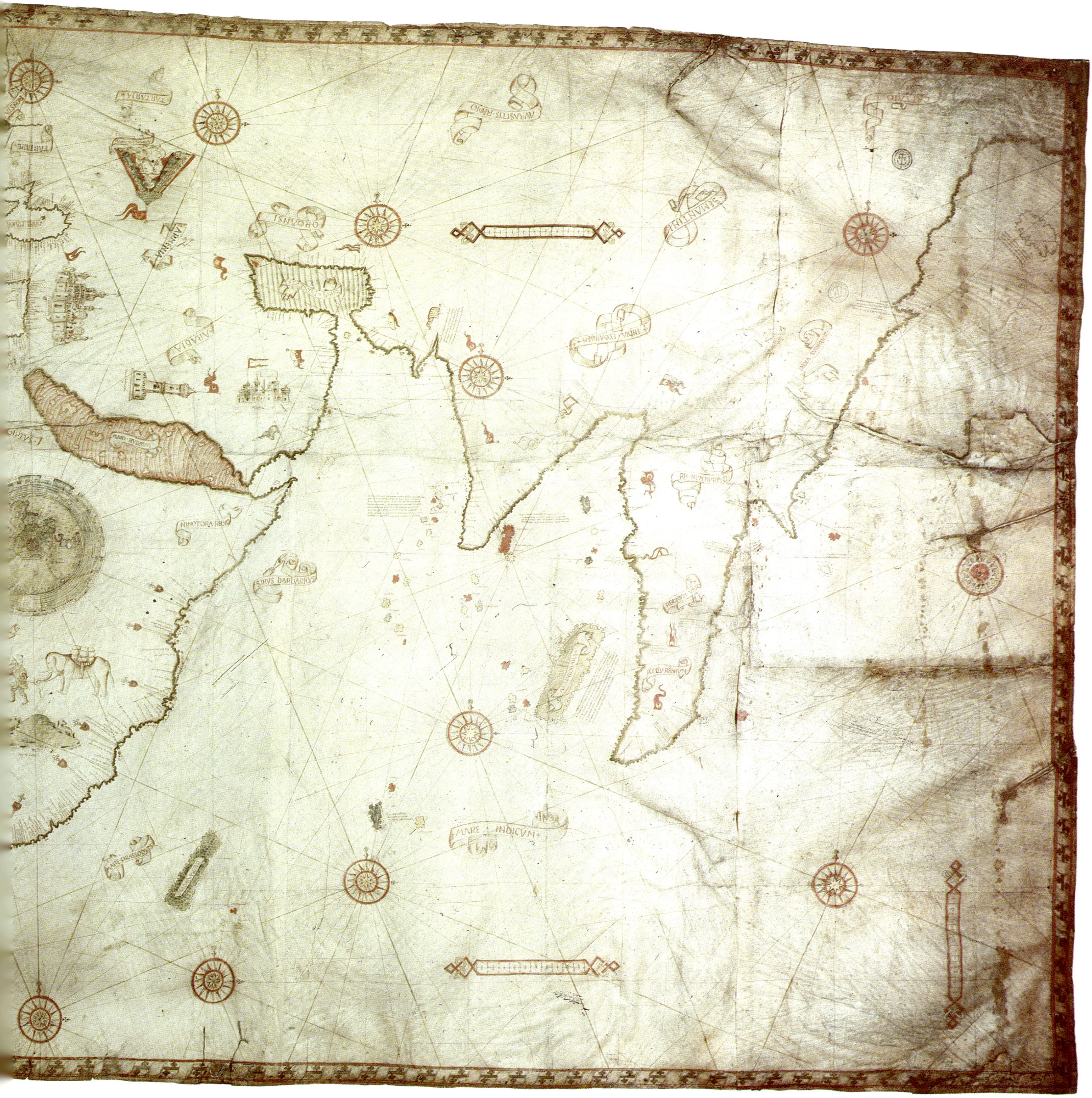
MARE + INDICVM

GIOVANNI MATTEO CONTARINI, WORLD MAP, FLORENCE, 1506

This is the oldest known printed map to show America. It is illustrated from the only known example, preserved at the British Library. Contarini used a novel conical projection, spreading the world in a fan shape to add information of new lands brought back by the first overseas explorers. The Contarini map appeared during the first decade of the sixteenth century when the results of Columbus's voyages to the west and Vasco da Gama's to the east were beginning to be recorded on maps. Accommodating the new discoveries was the challenge and dilemma of cartographers, and Contarini resolved it well. This world map represents the earliest attempt to bring the Far East and Far West into relationship to one another. Most maps of the period such as those of Caveri and Waldseemüller (Plates 13 and 16) show the eastern and western discoveries at the extreme right and extreme left and make no attempt to indicate their connection.

The mapmaker, about whom little is known, was probably a member of the distinguished Contarini family of Venice. Despite the fact that on the map he refers to himself as "famed in the Ptolemaean art," no other maps or charts by him have been discovered. The engraver, Francesco Rosselli, had been in the map trade since the early 1490s when he engraved Martellus's world map (Plate 5).

While the map seems puzzling at first, a closer look reveals the shape of the world as widely imagined by Europeans at the time of its publication. The three active European maritime nations, Spain, Portugal, and England, represented by Columbus, the Corte-Reals, and Cabot, all presumed that the new lands discovered in the West Indies and North America were in and around an extreme easterly promontory of the Asian continent. This is just how these lands appear on Contarini's map.

Off the eastern tip of that promontory an inscription refers to the discovery of Newfoundland by the Portuguese, a reference to the Corte-Real expeditions in 1500 and 1501. The legend at the West Indies reads, "the islands which Master Christopher Columbus discovered at the instance of the most serene King of Spain." As no coastline west of Cuba is marked, Contarini either was unaware of or chose to ignore Vespucci's alleged voyage to Florida in 1497 and Cabot's possible southern explorations on his second voyage of 1498.

The delineation of Africa is greatly improved, and India appears for the first time as a peninsula with Calicut named, reflecting the voyages of Vasco da Gama and of Cabral. Ceylon is considerably corrected, but the rest of the Asian cartography is essentially drawn from Ptolemy with some Marco Polo place-names.

Although the map shows some similarities with several early Portuguese manuscripts, it does not follow any of them closely. Contarini either synthesized a number of sources or employed a composite map that has not survived. His American place-names are found on the La Cosa, Cantino, or Caveri manuscript maps, and he included information found on each of them.

This cartographic treasure, unknown to scholarship until its discovery in 1922, gives a true reflection of cosmographical thought at the time Columbus's career was coming to a close. Perhaps its most charming feature is the legend off the east coast of Asia indicating that Contarini, along with Columbus in 1506, the year of the Admiral's death, believed that the great explorer had reached the coast of Asia. It reads: *Christopher Columbus, Viceroy of Spain, sailing westwards, reached the Spanish islands after many hardships and dangers. Weighing anchor thence he sailed to the province called Ciamba (the "Champa" of Marco Polo, known today as Indo-China). Afterwards he betook himself to this place which, as Christopher himself, that most diligent investigator of maritime things, asserts, holds a great store of gold.*

References: Fite & Freeman 1926, no. 6; Ganong 1964, 40-41; Hoffman 1961, 58-59; Layng (comp.) 1956, no. 18; Shirley 1983, no. 24; Sprent 1926; Williamson 1962, 141-42, 301-06; Wroth 1944, no. 23; 1970, no.9.

PLATE 14 (detail)
Giovanni Matteo Contarini
The West Indies, Japan,
and mainland Asia

CIRCIUS
SEPTE MTRIO
TANGVTP ROVINCIAMAGNA
PROVINCIA
KATHAY
HANCTERRAM
INVENERENAV
TELVSITANOR
REGIS
HEC SVNT INSVLE
CHRISTOTHORVS
AD IN STANTI
MI HISPANIE
TERRA
DECVBA
INSVLA
HISPANI
TOTVM MARETSTV
HORVS COLVMBVSVCEREX
OCCIDETE VERSVSNAVE
MVLTOS LABORES ET PERI
TAD INSVLAS HISPANAS
S NAVIGAVIT AD PROVI
MBA POSTEA COTVLI
ATOR ASSERIT
XIMAM
VINOCTIALIS

FAVONIVS
CIRCVLVS EQVINOCTIALIS
CIRCVLVS CANCRI
TERRA S CRVCIS
LIBOAVSTER

PLATE 14
Giovanni Matteo Contarini (d. 1507)
World map Florence, 1506

Copperplate engraving on paper,
17 x 25 in. (420 x 630 mm.)
Location:
The British Library/Map Library, London

UNIVERSALIOR EX RECENTIBUS CONFECTA

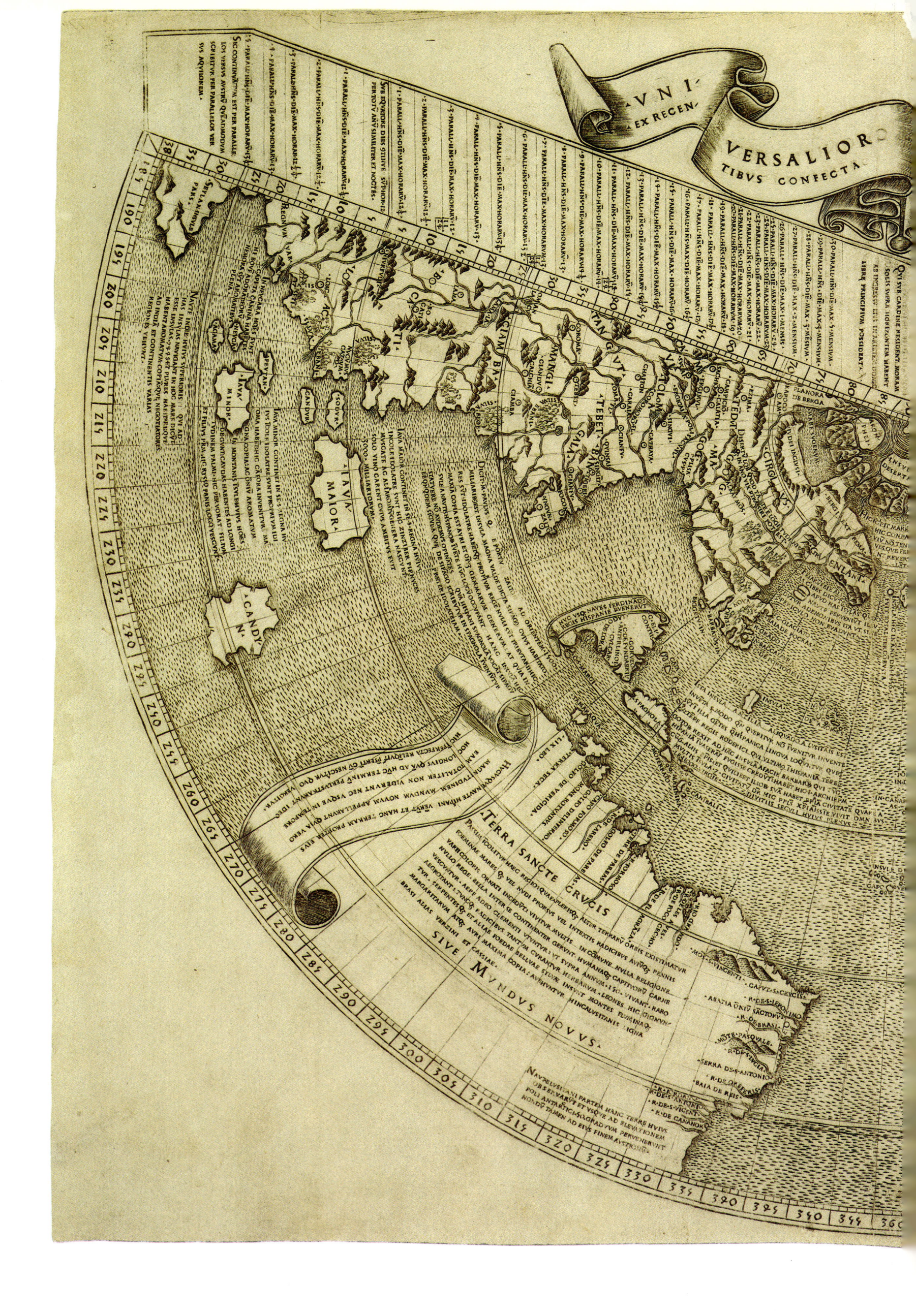

Early 16th-century world map (eastern hemisphere portion showing Asia, parts of the New World — "Mundus Novus" / "Terra Sancte Crucis" — and islands including Java Maior, Java Minor, Candyn, and Zipangri).

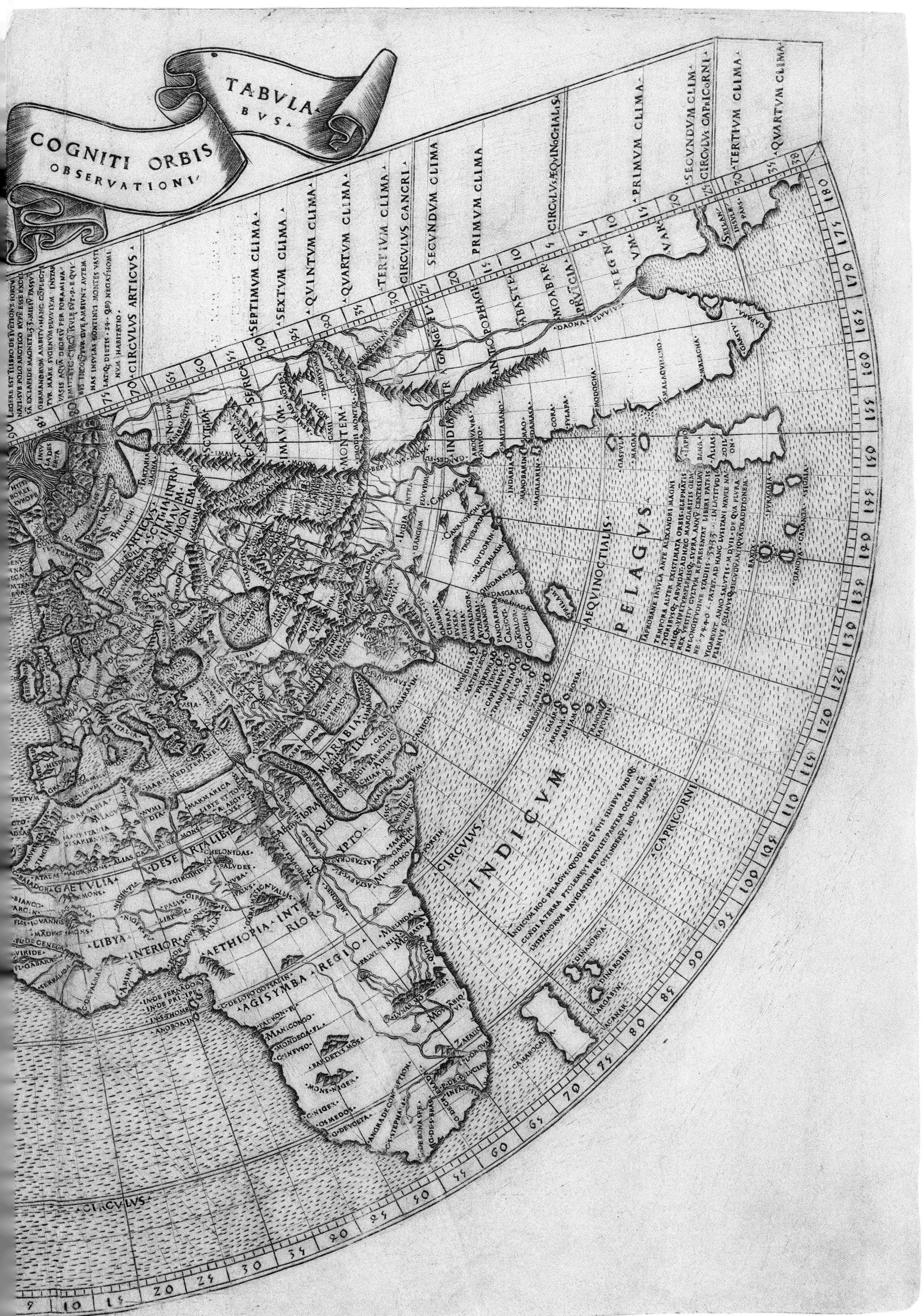

PLATE 15
Johannes Ruysch (d. 1533)
"Enlarged Map of the Known World Drawn from Recent Discoveries"
Rome, 1507

Copperplate engraving on paper,
16 x 21.5 in. (405 x 535 mm.)
Location:
Arthur Holzheimer collection

JOHANNES RUYSCH, WORLD MAP, ROME, 1507

Ptolemy's *Geography* was the most sought-after atlas during the early stage of the Age of Discovery and was reissued repeatedly. The first edition to be published after Columbus's voyages appeared in Rome in 1507. Some copies contained a revolutionary new world map compiled by Johannes Ruysch, a Dutchman living in Germany about whom little is known. When the very popular 1507 Rome edition of Ptolemy was succeeded by another edition the following year, the title page announced the inclusion of this new map of the world that displayed the lands discovered by European navigators.

An Italian monk, Marcus Beneventanus, wrote a description of this map for the 1508 edition. He stated that Ruysch had told him of sailing from the south of England to 53 degrees north latitude, then westward along that parallel where, bearing a little northward, he had observed many islands. Although it is not believed that Ruysch accompanied John Cabot, who reached North America before 1500, it is widely accepted that around that year Ruysch made a voyage with Bristol seamen to the great fishing banks off Newfoundland.

Information on America reached Europe from either Spanish, English, or Portuguese sources. The first maps issued in the early years of this new era often revealed the derivation of their place-names by the language source used. Ruysch employed Portuguese and had Lusitanian delineations as well. Although his conical, fan-shaped projection is similar to Contarini's (Plate 14), marked differences can be seen in the two cartographers' maps.

Both show that their information came from the Portuguese, but a comparison of the North Atlantic, where Greenland, Labrador, and Newfoundland had to be resolved, reveals that they treated the contours differently, particularly on the Arctic coastline. Contarini's delineation would permit a northwest passage, whereas Ruysch's would not. Differences in shapes also appear in their treatment of the West Indies and South America. Donald McGuirk's recent analysis of Ruysch's Cuba reveals that the island's delineation was changed during the map's production.

On the continent of South America, called *Mundus Novus,* Ruysch includes a legend discussing manners and customs of the natives. He mentions their cannibalism, good health, and longevity (150 years!). He writes of wild beasts and monsters as well as pearls and gold. Another inscription informs us that Portuguese navigators had sailed as far as 50 degrees south latitude without seeing the end of the continent. Where he felt obliged to draw a scroll on the west coast of South America, his Latin inscription explains, "as far as this, Spanish navigators have come, and they have called this land, on account of its greatness, the New World. In as much as they have not wholly explored it... further than the present termination, it must remain thus imperfectly delineated until it is known in what direction it extends."

Unlike mapmakers who came before and after him, Ruysch did not include "Cipango" (Japan). Continuing his candor, another of his legends reads: "M. Polo says that...there is a very large island called Cipango, whose inhabitants worship idols and have their own king.... They have a great abundance of gold and all kinds of gems. But as the islands discovered by the Spaniards occupy this spot, we do not dare to locate this island here, being of the opinion that what the Spaniards call Spagnola is really Cipango, since the things that are described as of Cipango are also found in Spagnola, besides the idolatry."

PLATE 15 (detail)
Johannes Ruysch
The West Indies

Toward the North Pole, Ruysch presents an innovative delineation including four fictitious islands in the Arctic Ocean. He shows a mountain of magnetism in an early attempt to depict magnetic deviation of the compass needle. His four Arctic islands were adopted sixty years later by Gerard Mercator on his famous world map (Plate 40).

Only two other printed maps showed America at this date, the Contarini and the Waldseemüller (Plates 14 and 16), and merely a single copy of each survives. Ruysch's world map, widely circulated because of its inclusion in a Ptolemy atlas, was one of the most influential of its time. More Europeans caught their first glimpse of the New World on Ruysch's fan-shaped *Tabula* than on any other map.

References: Fite & Freeman 1926, no. 9; Ganong 1964, 36ff; Harrisse 1892, 449-52; Hoffman 1961, 64-67; Layng (comp.) 1956, no. 5; Lowery 1912, no. 5; McGuirk 1986, 40-41; 1989, (a) & (b); Nordenskiöld [1889] 1961, 63-67; Shirley 1983, no. 25; Swan 1951; Woodward (ed.) 1987, 195-56; Wroth 1970, no. 12.

MARTIN WALDSEEMULLER, WORLD MAP, STRASSBURG, 1507

Just after 1500, at German-speaking St. Dié in the Rhineland, a school of cosmography and cartography was formed under the sponsorship of the Duke of Lorraine. The principals were the Duke's secretary, Canon Walter Ludd; his printer, Mathias Ringmann; and most importantly, cosmographer Martin Waldseemüller. The members of this *Gymnasium Vosagense* were intensely interested in news of overseas discoveries reported from Portuguese and Spanish sources. They were attempting to accommodate the newly described lands onto traditional maps of the world.

Waldseemüller prepared the most important charts published in Europe during the early decades of the Great Discoveries. This enormous world map of 1507 was the first printed version of a series called by modern scholars the Lusitano-Germanic group (Plates 11 and 13). These maps carried news of Hispanic overseas achievements from Portugal to the central European centers of cosmographical activity at Nuremberg, St. Dié, and Vienna. From there, the information was disseminated throughout trans–Alpine Europe.

Waldseemüller started with a Ptolemaic world delineation and revised it with information from Marco Polo's thirteenth-century travels in Asia, including the fabled Cathay and Cipangu (China and Japan) with their glittering riches. The large wall map by Martellus (Plate 5), which incorporated new Portuguese information on the African coast and had appeared just before Columbus's first voyage, was another Waldseemüller source. The continental configurations of the Old World and the pseudo-cordiform projections were similar in both productions.

The New World and the updated East Indies are clearly derived from the Cantino and Caveri charts (Plates 11 and 13). The delineation of Cuba and Hispaniola and their relationship to an incomplete coastline of the adjacent mainland are taken from these Portuguese sources. Columbus had visited Hispaniola, Cuba, Jamaica, Puerto Rico, and the Virgin Islands, which all appear prominently. Although his information is incorporated, it remains puzzling that the Lesser Antilles south of the Virgins and down to Trinidad are not shown, since they also were sighted during his first three voyages.

Waldseemüller projects a most important concept not found on previous maps, and surprisingly not on many maps that followed. This is the clear cartographic statement of the geographic independence of North America from Asia. The New World appears with a western coastline, indicating for the first time the existence of another sea separating North America and Asia, with Japan lying between. This concept is shown on the main map and on the New World inset at top right. Waldseemüller is not as consistent regarding the supposed strait at Central America that divides North and South America and connects the two oceans. He included it on the large map but not on the inset. This nonexistent strait was Columbus's chief target on his final voyage. His inability to discover a passage by sea to Cathay remained his greatest disappointment.

In 1504 the *Gymnasium* published a pamphlet containing a letter from Amerigo Vespucci about his Third Voyage to *Novus Orbis* (South America). Three years later, Waldseemüller recommended, in his "Introduction to Cosmography" written to accompany this large world map, that the newly discovered continent be called *America* in Vespucci's honor. It is unlikely he intended this to apply to North America, whose size and shape was still a mystery.

The *Introduction* has always been known, and from it scholars were aware that the map had been made. However, no copy had come to light until this one was discovered in 1901. It remains the only surviving example from a large printing. On the South American land mass the name "America" appears for the first time on a dated map. With the *Introduction* and this map, the influential Waldseemüller effectively named the New World not after the man who discovered it, but in honor of another voyager more gifted in the art of public relations.

References: Bagrow/Karrow 1990, no. 80:2.1; Brown 1950, 15657; Cortesão 1969-71, 2:108ff; Cumming 1962, pl. 1; Fischer & von Wieser 1903; Fite & Freeman 1926, no. 8; Layng (comp.) 1956, no. 19; Schilder 1986-88, 2:11-16; Shirley 1983, no. 26; Williamson 1962, 319-20; Wroth 1944, no. 24; 1970, no. 10.

PLATE 16 (detail)
Martin Waldseemüller
Amerigo Vespucci with North and South America connected

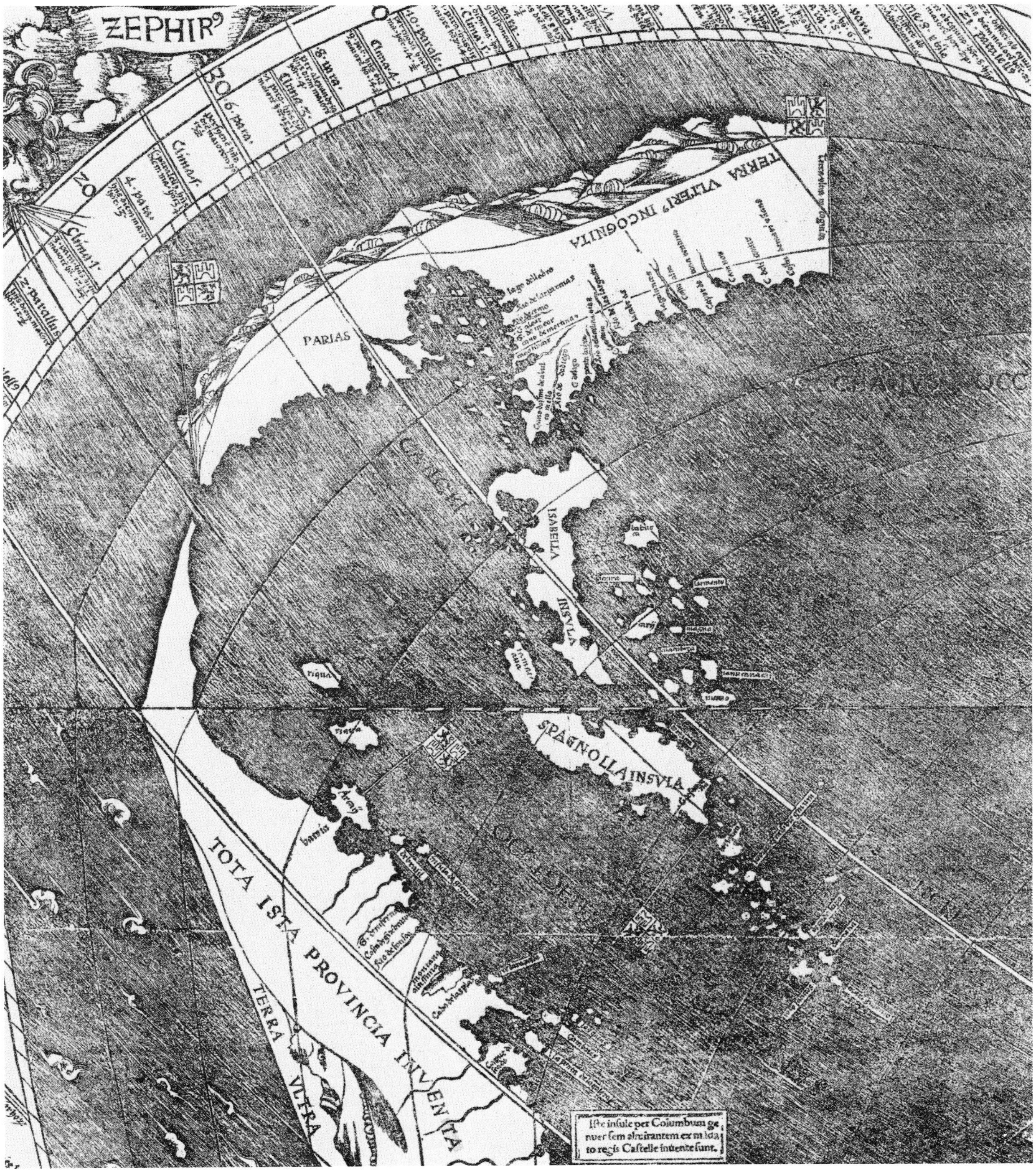

PLATE 16 (detail)
Martin Waldseemüller
The New World with strait separating North and South America

VNIVERSALIS   COSMOGRAPHIA   SECVNDVM PTHOLOMAEI TRA   DITIONEM

PLATE 16
Martin Waldseemüller (1470–1521)
Universalis Cosmographia Secundum Ptholomaei Traditionem
Et Americi Vespucii Aliorūque Lustrationes, *Strassburg, 1507*
Woodcut on paper, 12 sheets designed to be joined,
53 x 94 in. (1320 x 2360 mm.)
Location:
Schloss Wolfegg, Württemberg (Germany)

FRANCESCO ROSSELLI, MARINE CHART & WORLD MAP, FLORENCE, CA. 1508

Francesco Rosselli, member of a distinguished Florentine family of artisans, was a cartographer, engraver, and map-dealer whose shop in Florence is the earliest map-selling establishment on record.

Cartography was a new, experimental art and science during the first years of the sixteenth century, and Rosselli was among its gifted innovators. The conical projection world map on which he collaborated with Contarini (Plate 14) is the first such engraving to have appeared. Both men are remembered principally for this production, but Rosselli also left to posterity the two small world maps shown here. These undated engravings were executed in 1507 or 1508 at either Florence or Venice. Both are rare, important documents known by only a few examples.

Another cartographic projection made its debut on Rosselli's oval planispheric world map (Plate 17B). The design was adopted and used by some of the century's most famous mapmakers: Bordone, Münster, Gastaldi, and Ortelius. Rosselli improved the configuration of North America while still displaying its attachment to Asia. Except for the new discoveries in the extreme northeast, the Asian delineation is mostly derived from the classical productions of Ptolemy with some additions from Marco Polo. The central meridian crosses Russia, East Africa, and Arabia, an inconvenient division of the world for following the new discoveries.

For the sea chart (Plate 17A), with its circle of sixteen compass roses radiating rhumb lines across the surface, Rosselli placed the prime meridian in the eastern Atlantic, helping to interpret Columbus's cosmographical concepts more clearly. On this chart it is easy to see why Columbus, sailing west of Cuba, believed the continental coastline he encountered was that of China.

Rosselli's maps record information about discoveries in North America by Columbus and Cabot that had not been shown earlier. They are important as representing the geographical concepts of Columbus's Fourth Voyage in 1502–03. Columbus designated many place-names in Central America when he sailed from Hispaniola to northern Honduras and along the coasts of Nicaragua and Costa Rica to eastern Panama. On the Rosselli maps, these names appear on the east coast of Asia along China and Indo-China, which, after all, is where Columbus thought he was.

The beautifully colored examples of Rosselli maps at Greenwich were long thought to be manuscripts. Because they were printed on vellum and the oval planisphere is richly illuminated, this misconception is understandable. Comparisons with the black-and-white engravings on paper at Florence revealed that the Greenwich pair, beneath the color, were identical. We have illustrated detail from the only other known copy of this engraving to demonstrate that place-names obliterated on the painted example by the ocean tint are clearly legible on the black-and-white engraving.

As with all known world maps of the first decade of the sixteenth century, Rosselli's delineations provide a key graphic link between pre- and post-Columbian cartography. Little by little the mysteries of the New World were being solved, and the Ocean Sea was giving up its secrets.

References: Bagrow/Skelton 1964, 368; Brown (comp.) 1952, no. 84; Harrisse 1900, 65-69; Hough 1980, no. 98; Howse & Sanderson 1973, no. 4; Nunn 1928; *Rosselli Oval Planisphere* 1982; Shirley 1983, nos. 28, 29.

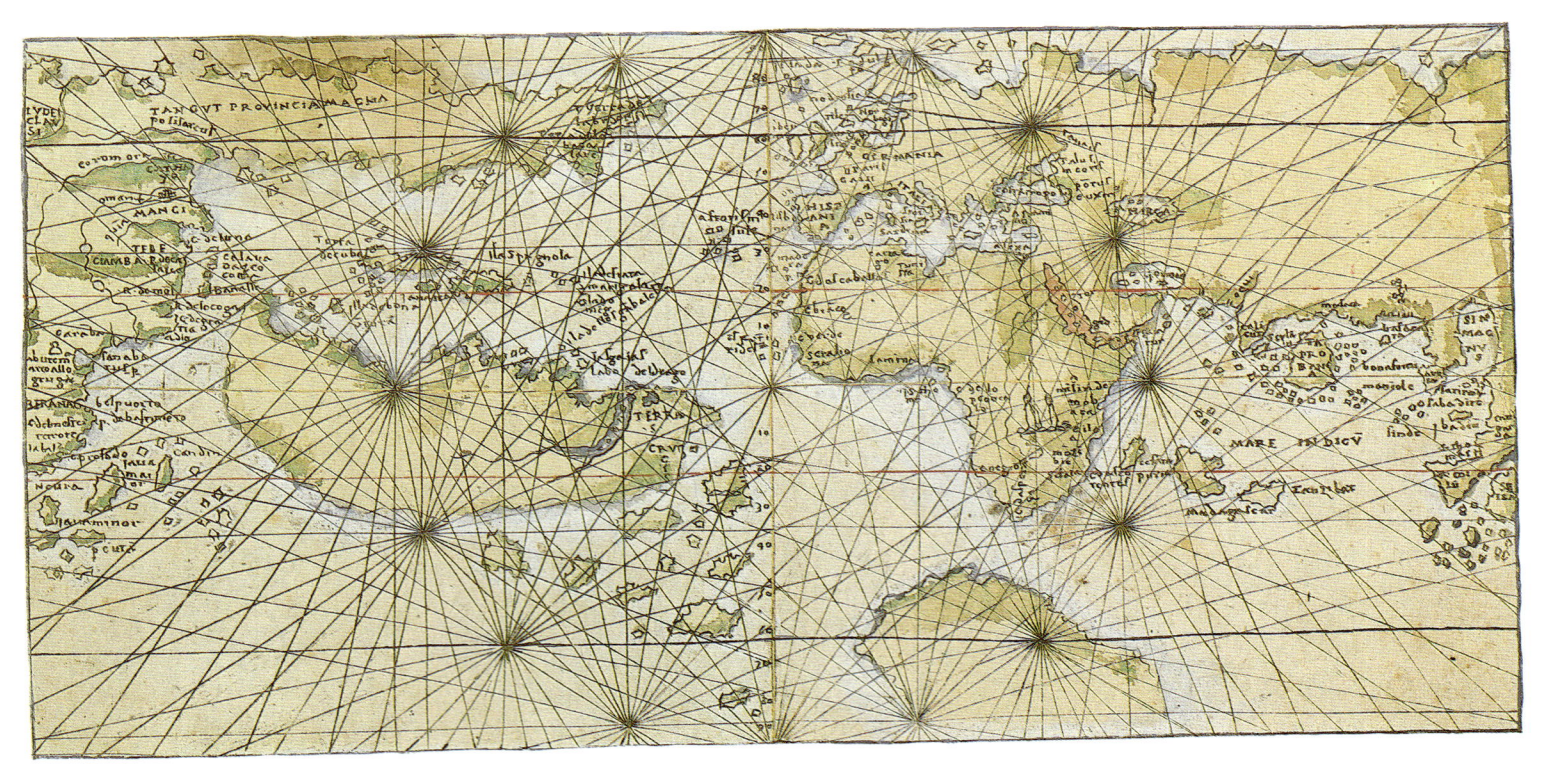

PLATE 17A
Francesco Rosselli (ca. 1445–1513)
Marine chart of the world
Florence, ca. 1508
Copperplate engraving on vellum, colored by hand, 7 x 13 in. (180 x 335 mm.)
Location:
National Maritime Museum, Greenwich

PLATE 17B
Francesco Rosselli (ca. 1445–1513)
Oval projection world map
Florence, ca. 1508

Copperplate engraving on vellum, colored by hand,
6 x 11 in. (145 x 285 mm.)
Location:
National Maritime Museum, Greenwich

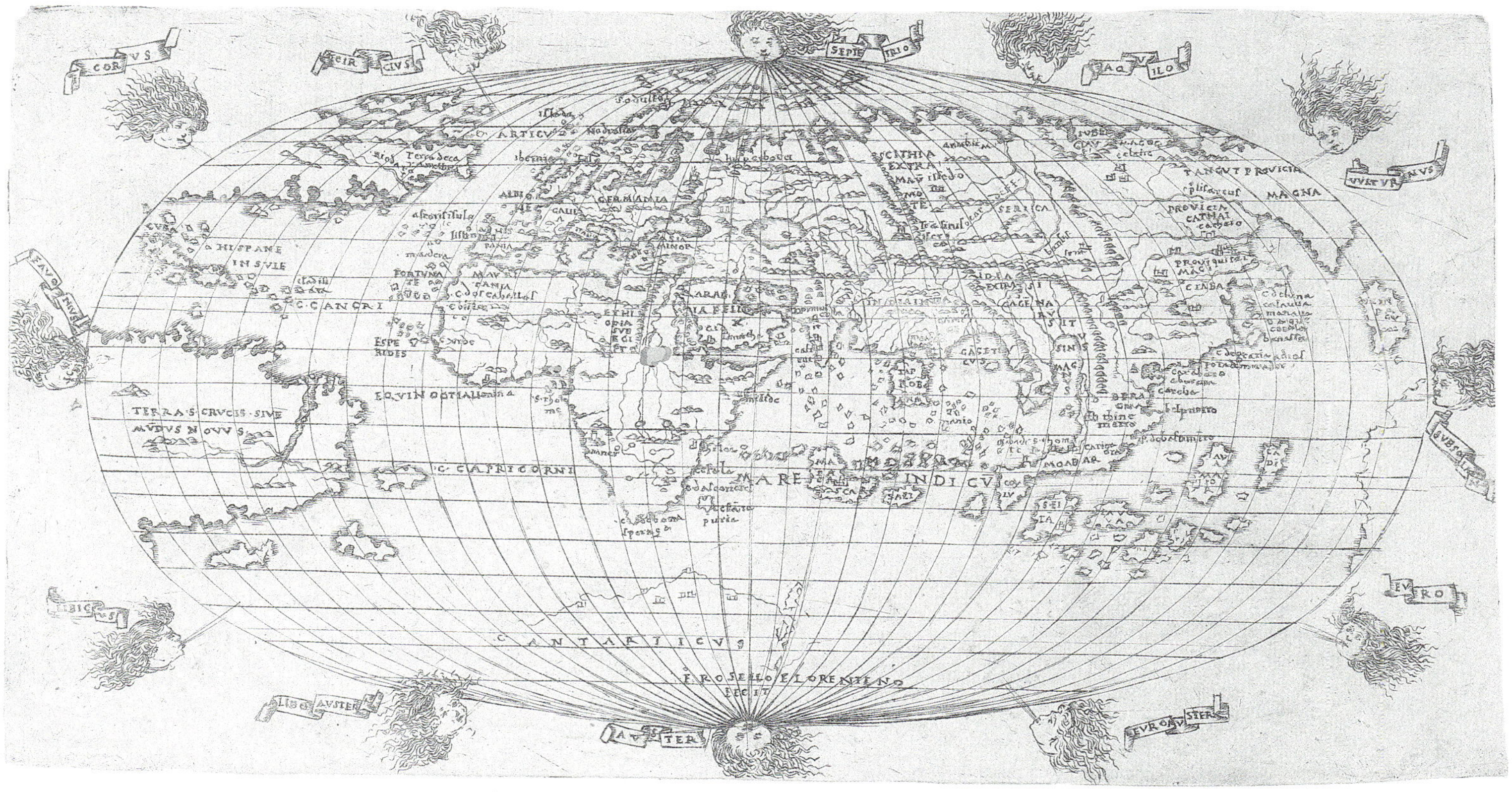

PLATE 17C
Francesco Rosselli
An uncolored example revealing obscured place-names
Location:
Arthur Holzheimer collection

57

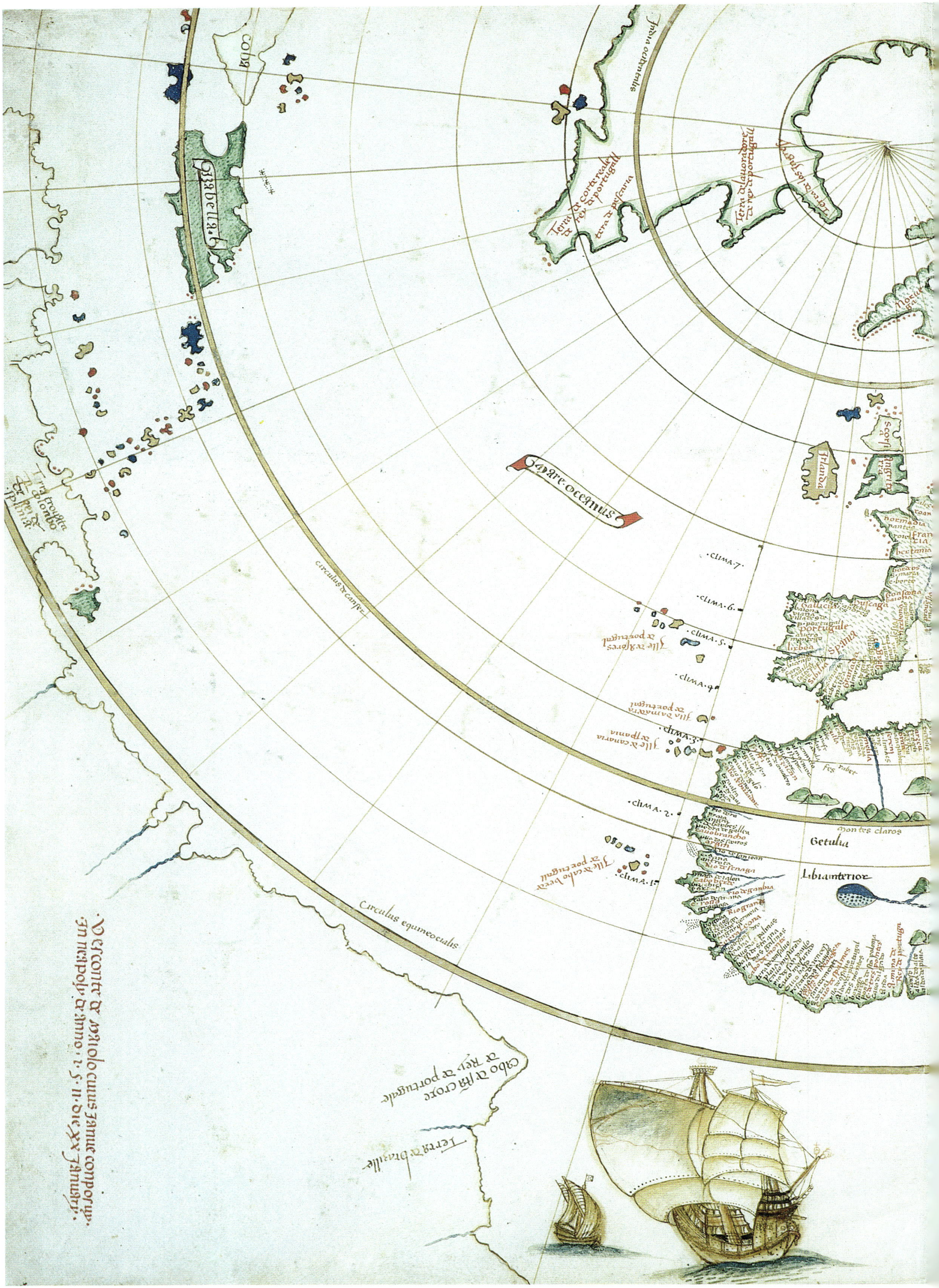
Circulus equinocialis
Getulia
Libia interior
Irlanda

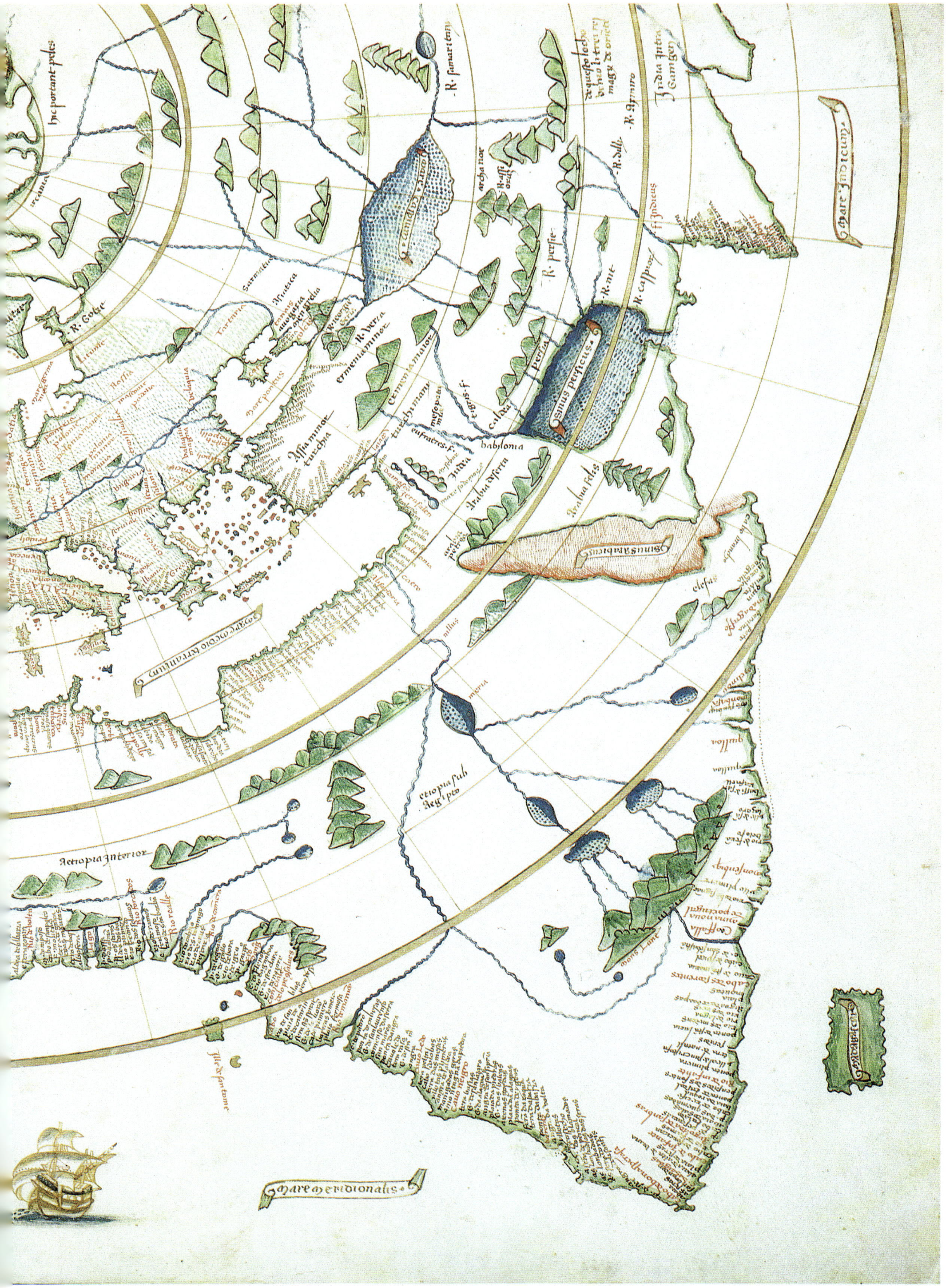

PLATE 18
Vesconte de Maggiolo (fl. 1504–51)
World map
Naples, 1511
Illuminated manuscript on vellum,
15 x 22 in. (390 x 560 mm.)
Location:
John Carter Brown Library,
Brown University, Providence

VESCONTE DE MAGGIOLO, WORLD MAP, NAPLES, 1511

Maggiolo was a prominent Genoese cartographer who produced important manuscript maps during the first half of the sixteenth century. From 1511 to 1518 he worked in Naples, where for a noble Corsican family he prepared the beautiful atlas that includes this map. It is signed and dated, "Vesconte de Maiolo, from Genoa composed it in Naples in the year 1511 on the 10th day of January."

Maggiolo founded a school of chartmakers for the Republic of Genoa. He had been summoned there by the doge of the city-state, who proclaimed that Maggiolo was "skilful in the making of nautical charts and other things requisite for navigation." About twenty of his portolan charts and atlases have survived. His descendants were mapmakers for more than a century after his death.

This map is drawn with a north polar projection that provides its distinctive fan shape. Its format resembles those of Contarini and Ruysch (Plates 14 and 15), which are derived from conical projections. Maggiolo has not attempted to display the full 360 degrees of the sphere; less than 200 degrees appear, leaving East Asia and the Ocean Sea incomplete. His interpretation of longitude within this projection continues the elongated east-west dimensions for the Mediterranean, Eurasian, and African areas seen on earlier maps. The farther south the area, the more pronounced this effect becomes.

The single Arctic and North Atlantic landmass at the top indicates that the location of the new discoveries was still thought to be in far northeastern Asia. Maggiolo's map shows a solid Eurasian continent running from Scandinavia around the North Pole, including Asia's arctic coast, to Newfoundland-Labrador and Greenland. On the extreme northeast promontory of North America, Maggiolo place-names include "land of the English," and "land of the Corte-Real and of the King of Portugal." Just westward the presence of the name *India Occidentalis* (West Indies) appears for perhaps the first time on a map.

South America is annotated, "lands found by Columbus for Spain" in Venezuela. Below the equator on the coast are the words, "Cape of the Holy Cross of the King of Portugal" and "land of Brazile." In the West Indies Cuba is named, but Hispaniola is called Isabella. The Lesser Antilles continue from the Virgin Islands south to Trinidad without the hiatus in some earlier maps. Indications of small islands appear in the area of the Bahamas and the Turks and Caicos.

Drawn and decorated for a noble patron, signed and dated by the cartographer, this elegant production is a significant representation of the world just after Columbus's final voyage, which concluded the first phase of the Age of Discovery.

References: Caraci 1937; Harrisse 1892, no. 83; Hoffman 1961, 61-63; Hough 1980, no. 101; JCB Annual Report 1919-31, 52; JCB Annual Report 1913, 19; 1929, 38-39; 1965, 13-14; Mapes 1963; Wroth 1970, no. 17.

PIETRO MARTYR D'ANGHIERA, MAP OF THE INDIES, SEVILLE, 1511

A handful of manuscript maps showing the New World were produced in Spain and Portugal during the early stages of the discovery period. They were not widely known, as such maps were considered state secrets and revealing them was punishable by death. Occasionally, as in the case of the Cantino map (Plate 11), a manuscript copy was smuggled out. In general, however, the important geographical information on such maps remained tightly controlled.

Against this background the woodcut illustrated here, which is the first printed Spanish map portraying America and the first separate map of the New World, raises as many questions as it answers. It was probably drawn by Andrea Morales, pilot and cartographer of the House of Trade at Seville, who was charged with construction of the royal map of the world's ports and harbors.

Peter Martyr, author of the book in which the map was issued, was the first historian of the New World. Like so many who played key roles in the early discovery period, Martyr was an Italian expatriate. Born outside Milan to a noble family, he was sent to Rome for religious training. After becoming a priest, he went to Spain in 1494, became tutor to the children of Ferdinand and Isabella, and was appointed to the powerful Council of the Indies.

His credentials for producing the first historical account of America were unusually strong. Martyr had the ear of the sovereigns and a position of influence, combined with an intense interest in the voyages of discovery. He also developed close personal friendships with many prominent explorers, including Columbus, Vespucci, Cabot, Magellan, Vasco da Gama, and Cortez. He received direct communications from these friends and other participants in the great drama.

It seems inconsistent that the author, a notable

in royal and scientific circles, would publish a map disclosing the location of the islands of the Indies at this critical period, when the monarchs wanted the information suppressed. Nevertheless, the map was produced, although no copies seem to have appeared in the first issue of the book. The quality of the delineation is still debated between those who feel it is remarkably accurate, considering its early date, and those who assert it to be grossly inadequate.

That the map's information is current is shown by the first appearance of Bermuda, discovered by Juan Bermudez in 1505, on his homeward passage from Hispaniola. However, the relationships suggested on the Ocean Sea of Bermuda, the Canary Islands, and the Antilles with Europe and America are very misleading. They imply a diminutive ocean and the possibility of a relatively brief crossing. This arrangement reflects the theory promoted by the Florentine cosmographer, Toscanelli (d. 1482), which fueled Columbus's desire to make his historic attempt to reach the Indies. The obvious mixing of scales contributes to the map's apparent inaccuracies. In addition to the foreshortened oceanic distances, Cuba and Hispaniola appear much larger than the South and Central American coastlines with which they are juxtaposed.

Another inconsistency is the absence of any of the Lesser Antilles except Trinidad and "Isla Verde." Columbus discovered Trinidad on his Third Voyage. However, the other islands of the Lesser Antilles from the Virgin Islands south to Guadaloupe and Dominica (which he called the Cannibal Islands), visited on his Second Voyage, are not shown on the Peter Martyr map. A printed note on the reverse side advises that those islands were omitted "to avoid confusion."

Despite its mystifying elements, this is still the best map of the New World from the period. It would be many years before an improved printed delineation would be available to the public.

References: Brown (comp.) 1952, no. 52; Hoffman 1961, 61-63; Layng (comp.) 1956, no. 32; Lowery 1912, no. 6; Quinn 1989; Skelton 1958, 59.

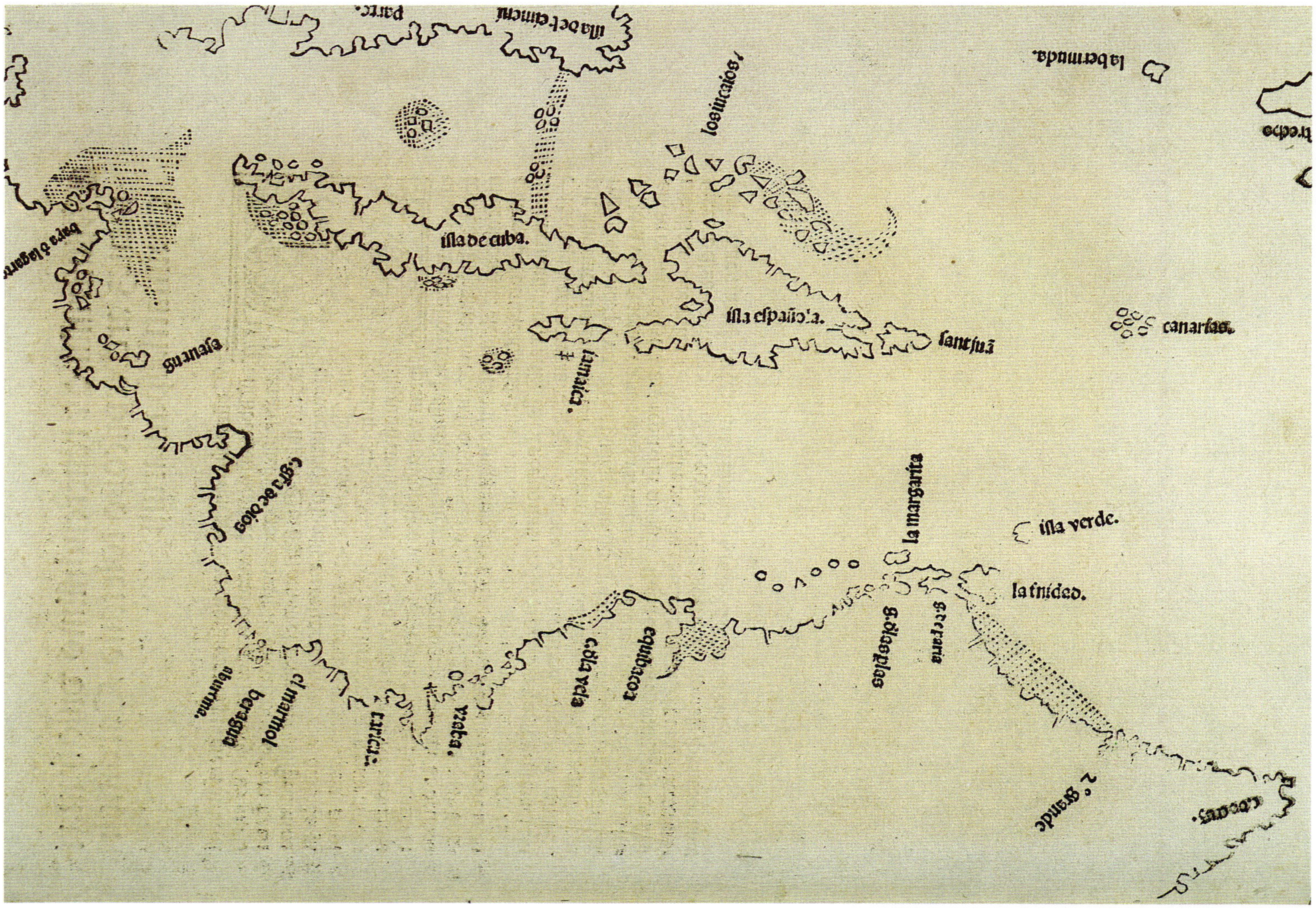

PLATE 19
Pietro Martyr d'Anghiera (1455–1526)
Map of the Indies
Seville, 1511

Woodcut on paper, 7 x 7 in. (190 x 180 mm.)
Location:
The Newberry Library, Chicago

PIRI RE'IS, CHART OF THE OCEAN SEA, GALLIPOLI, 1513

During a naval campaign against Venice in 1501, a Turkish fleet captured a Spanish ship in the western Mediterranean. One of the prisoners taken had earlier made three voyages to the West Indies with Columbus and carried with him a set of Columbus's American charts. In this fortuitous manner Kemal Re'is, the famous Turkish admiral, acquired maps of great importance showing a newly discovered part of the world.

Piri Re'is, nephew of Kemal, was born in Gallipoli on the shore of the Dardanelles in 1470. Piri also became an admiral and is remembered as a scholar of navigational science and an accomplished linguist. He produced charts, an important book on navigation, and a superb map of the world, which employed the Columbus maps taken by his uncle's sailors. Although fragmentary, this work and the Zorzi sketches (Plate 12) are the only world maps with a direct Columbus delineation for part of America.

The map found its way to Suleiman the Magnificent's Topkapi Palace where it remained undetected for four centuries. In 1929 this fragment was discovered when the palace was being converted to a national museum. Many lengthy notes in Turkish appear on the map, including geographical descriptions and detailed information on the sources of the delineation. There are references to the voyage of St. Brendan, the legendary Irish monk who in the sixth century supposedly discovered an island in the North Atlantic called the "Promised Land of the Saints." Long sought by sailors, St. Brendan's island was widely believed to exist in Columbus's time and appeared in some form and location on most early European maps.

A long passage describes Columbus's First Voyage experiences, from initial difficulties in obtaining sponsorship to encounters with the natives. Piri Re'is specifically mentions his use of the West Indies charts drawn by Columbus. He also refers to information from Portuguese and Arabic sources that proved important in developing his delineation of Africa and Asia.

The style of the map is European although the lengthy commentary is written in Turkish. Piri comments that no one in Turkey had ever seen such a map. Presumably he referred to both the novelty of its delineation and the profuse depictions of people and animals that violated the customary Islamic prohibition against portraying living objects in artworks. The map was not only unusual in Turkey, but few people in any country, including Spain and Portugal, had access to a chart of the world incorporating the new discoveries.

The coastline of northeastern South America indicates that information came from Ojeda, Vespucci, or one of their companions. The West Indies are poorly drawn and difficult to recognize, but Guadaloupe and the islands immediately adjacent in the Lesser Antilles are remarkably accurate. For these Piri Re'is no doubt had a Columbus drawing. This unusual chart with its complicated and fascinating history includes the only surviving delineation by Columbus of his discoveries.

References: Akcura 1935; George 1969, 60-62; Kahle 1933a; Layng (comp.) 1956, no. 40; Mollat du Jourdin 1984, no. 28; Morison 1942, 2:71-72, 77.

PLATE 20
Piri Re'is (d. 1554)
Chart of the Ocean Sea
Gallipoli (Turkey), 1513
Illuminated manuscript on vellum,
35 x 25 in. (900 x 630 mm.)
Location:
Topkapi Saray Museum, Istanbul

MARTIN WALDSEEMULLER, *TERRE NOVE*, STRASSBURG, 1513

Waldseemüller and his colleagues in the *Gymnasium* at St. Dié began making maps for their new edition of Ptolemy's classical geography as early as 1505 (Plate 16). The wood blocks were completed by 1506–07, but publication was deferred until 1513. Except for this unexplained delay, *Terre Nove* would have been the first printed map of America.

Published in substantial quantities and widely distributed, Waldseemüller's maps had considerable influence. His sources were Portuguese charts such as the "Cantino" and Caveri (Plates 11 and 13). Spanish reports of the same period, as reflected in Peter Martyr's map of 1511 (Plate 19), provided a different configuration, as clearly illustrated by comparing Cuba on this work and the Peter Martyr map.

Early mapmakers' problems are often revealed by comparing treatment of the same area on different maps. For example, when Waldseemüller changed from the truncated, heart-shaped world that encompassed America in its western sector (Plate 16) to this plane chart of the new discoveries in the "western ocean," the latitudes of the West Indies became confused. Cuba appears directly on the Tropic of Cancer on the world map, while in the *Terre Nove* it is north of that line.

In addition, although Hispaniola is south of the Tropic on the world map, it lies north of it on the America chart. Hispaniola is delineated with a southwest-to-northeast axis on the world map, but revised on *Terre Nove* to east-west. If these small distinctions appear trivial, imagine the chaos they create for today's researchers who attempt to determine which small Bahamian island actually was the first landfall of Columbus.

In 1513, the *Terre Nove* was issued along with nineteen other "modern" maps in the most important sixteenth-century edition of Ptolemy's *Geography*. By then Waldseemüller had realized his earlier injustice in naming the entire region for Amerigo Vespucci while overlooking Columbus's name. The word *America* was now deleted and a two-line commentary added just below the equator indicating that Columbus was the original discoverer, sailing in the name of the King of Spain. In the introduction to the volume, Waldseemüller identified "The Admiral" as the source of information for his delineation of the New World. During this period no other sea captain was called by that title. Consequently, *Terre Nove*, with its Columbus information, is widely known as "The Admiral's Map."

References: Bagrow/Karrow 1990, 80:33; Cortesão 1969-71, 1:114-20; Cumming 1988, pl. 3; Nordenskiöld [1889] 1961, pl. 36; Skelton 1958, ch. 3; 1966.

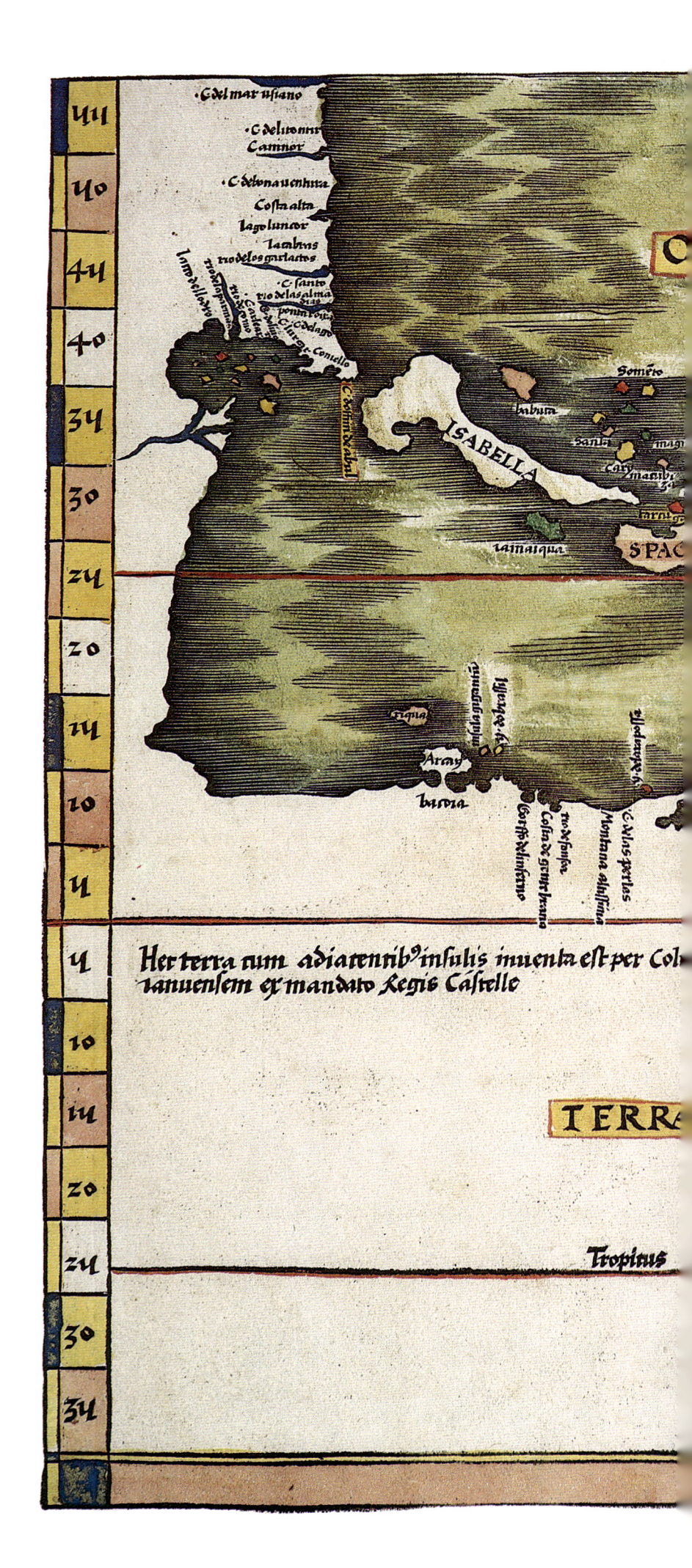

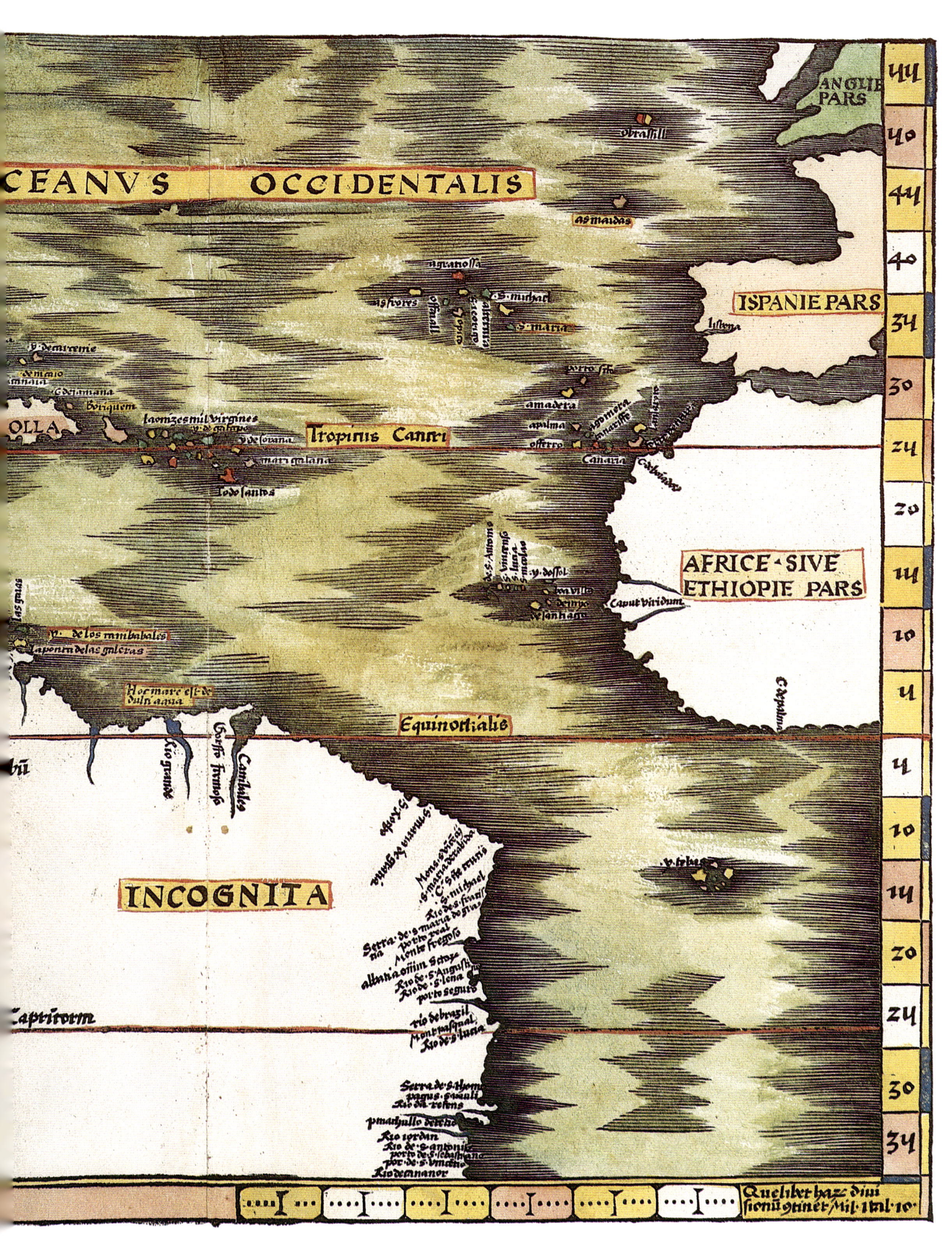

PLATE 21
*Martin Waldseemüller
(1470–1521)*
Terre Nove
Strassburg, 1513
Woodcut, colored by hand,
15 x 19 in. (380 x 490 mm.)
Location:
Ohio State University,
Columbus

LOPO HOMEM WITH PEDRO REINEL, NORTHERN INDIAN OCEAN, PORTUGAL, CA. 1519

While Columbus, Cabot, Vespucci, and the Corte-Real brothers were exploring the New World; Diaz, Vasco da Gama, and Albuquerque had brought the Portuguese flag and influence past the Cape of Good Hope to India and beyond. During this early period, the Portuguese controlled the African route to the Indies and their riches. Meanwhile the Spanish, English, and French competed fiercely to find a route around or through America, the barrier that blocked their way to those same Indies.

This spectacular chart illustrating the progress of European knowledge of Arabia, India, and the gateway to the Spice Islands is part of a remarkable atlas. King Manuel of Portugal reportedly commissioned the charts, probably for François I of France. They were then acquired by the French national library at the end of the nineteenth century. The manuscript is referred to as the *Miller Atlas* after the name of its last private owner. Portugal's best mapmakers, Master of Sea Charts Lopo Homem and the father-and-son team of Pedro and Jorge Reinel, produced the maps, which were complemented by the rich illumination of artist Gregorio Lopes. The atlas covers the entire known world, although the sheets for Africa unfortunately have not survived.

The first Portuguese voyages to the Indian Ocean took place between 1489 and 1515. Much of the new information from the Orient was first shown on this Indian chart. The place-names from Aden eastward along the south coast of Arabia and down the west coast of India indicate Portuguese landfalls. More prominent towns are represented by beautifully rendered, stylized citadels. At Mecca a miniature depicts the Kaaba, built in the seventh century and standing today as the holiest place in Islam.

This is one of the earliest charts to portray somewhat accurately the Ganges Delta and the Bay of Bengal and to give proper alignment to the Burmese coast. At lower center the huge archipelago of multicolored islands was refined in later maps into the Laccadives and Maldives. Finally, after the grotesque distortions of earlier works, Ceylon and the northwest part of Sumatra (Taprobana) are located comparatively correctly.

The chart's decoration is enriched by well-rendered ships. The two at lower left are Portuguese. Others flying the Islamic crescent are from Admiral Piri Re'is's Indian Ocean squadron of the Ottoman fleet, and the large twin-ruddered craft east of India are Chinese vessels. North of India the finely drawn elephants, rhinoceros, and lion provide images of the exotic fauna that became symbolic of southern Asia. An equestrian warrior gallops across the Horn of Africa, a pair of camels wind through Arabia, and in central India two fierce natives prepare for combat. Beautiful flora appropriately selected for each region, Portuguese flags and coats-of-arms, and the magnificent compass rose complete the composition. This is one of the few surviving charts whose artistic beauty matches its importance as a primary document in the history of exploration.

References: Cortesão & da Mota 1960 1:55-61, pl. 19; George 1969, 117; Lach 1965, 1:221; Mollat du Jourdin 1984, no. 31.

LOPO HOMEM WITH PEDRO REINEL, EAST INDIES, PORTUGAL, CA. 1519

This historic chart of the East Indies and the previous map (Plate 22) are from the same sumptuous maritime manuscript atlas produced for the King of Portugal. Beginning with the Columbus era, maps showing improvements in geographical knowledge contributed by returning explorers depict a continually revised outline of America. The true objective of those early voyages, however, was a westward sea route to the Spice Islands of the Orient seen here.

This Homem-Reinel chart is the earliest known cartographic survey of the Moluccas group. It is also the first relatively accurate map of Sumatra (still called Taprobana), with a recognizable Strait of Malacca. Except for the absence of the Gulf of Siam, Southeast Asia is comparatively well executed. The Mekong Delta is visible, as is Mergui Archipelago west of lower Burma. The south China Sea and Indonesia are represented as a Portuguese waterway, indicated by the circle of flags surrounding it. Two Lusitanian and five Islamic ships emphasize the continuing maritime conflict among European nations over the Spice Islands and trading centers of Malacca Straits.

This unique example of Portuguese cartography from the discovery period superimposes new information of geographical discoveries upon Ptolemaic geography of the classical period. The "climate zones" inherited from the ancients are displayed, while to the right, a solid landmass represents the eastern coastline of Ptolemy's *Sinus Magnus* (Great Gulf). Ptolemy's world map (Plate 1), clearly illustrates the Asian concepts of Greek times from which the *Sinus Magnus* coast is derived.

Malacca, captured by the Portuguese in 1511, served as the principal trading and market center for Oriental spices and silk. Interestingly, it appears here twice, probably because returning voyagers, stressing its great importance, may have called it by slightly different names. The entrepôt is symbolized by great white cities guarded by five-story towers, one labeled Malaqua, the other Mabaqua. To the northwest is the golden city of Pegu, an ancient metropolis in southern Burma, whose history dates to the sixth century A.D. Early European travelers were astonished by Pegu's many pagodas, huge golden shrine, and the colossal tenth-century reclining statue of Buddha. The figure, measuring 181 feet (54 meters) long, was called the most impressive monument of southern Buddhism.

This early important sea chart, drawn and decorated for a royal patron, reveals much new information on the Far East, dramatically portrayed by the juxtaposition of ancient and modern knowledge.

References: Cortesão & da Mota 1960 1:55-61, pl. 19; George 1969, 117; Lach 1965, 1:221; Mollat du Jourdin 1984, no. 31.

Map Labels

- DESSERTO
- PAROPANISAD
- ELEVFRATEZ
- SIANA
- PERSIA SOPIRIO
- ARABIA DESERTA
- MARE PERSICO
- PSIDIA
- DROTAE
- CARMANA
- INDIA
- CABAYA
- ABA REGIO
- MECHA
- CIRCVL9
- REGNV
- BETINVS F
- MARTIMOTES
- ADEM REGIO
- ARABIA FELIX
- ORMANVS
- METRAONS
- VBRVM MARE
- BARBATO
- COGOTORA
- SINVS ARABICVS
- MARE INDICVM
- PTOLEMOTES ETI
- ELEPHAS MONS
- OPIA
- CIR CVL VS

PLATE 22
Lopo Homem
(fl. 1497–1572)
with *Pedro Reinel*
(fl. 1485–1535 and 1518–72)
Northern Indian Ocean
Portugal, ca. 1519
*Illuminated manuscript
on vellum, 16.5 x 23.5 in.
(415 x 590 mm.)*
Location:
Bibliothèque
Nationale, Paris

PEGVI REGIO   MAE ANDRVS MON

MABAQVA CIVITAS

MARE INDI CVM

SVLE MOCALO

M LAQVA

GAMISPOLIS

Ante et post tapropa nam multitudo est Insularum quas di cunt esse numero mi lessimo trecentesimo septuagesimo octauo quarum tamen nomi na traduntur haec sui

MACAR INSVLE

TAPROBANA INSVLA

JAVA MAJOR INSVLA
JAVA MINOR

CANDIN INSVLA

PLATE 23
*Lopo Homem
(fl. 1497–1572),
with Pedro Reinel
(fl. 1485–1535 and
1518–72)*
East Indies
Portugal, ca. 1519
*Illuminated manuscript
on vellum, 16 x 20 in.
(415 x 590 mm.)*
Location:
Bibliothèque
Nationale, Paris

PART III

Filling in the Features of the Earth

Columbus, one of the greatest navigators of all time, was also among the worst colonial administrators. Although he discovered many more Caribbean islands and explored further along the South and Central American coasts, he had to be removed as governor of the new lands. Physically and emotionally exhausted by his epoch-making adventures, Columbus died in 1506 at the age of fifty-five.

At about the same time, John Cabot also crossed the Ocean Sea, sailing for Henry VII of England. Cabot, another Italian in the service of a foreign sovereign, was attempting to discover a shorter route to the Indies along the northern latitudes. If he succeeded, he might lift England from being last and farthest from the spice trade to being the first and closest. Cabot landed in maritime Canada in 1497 but perished at sea the following year attempting a second voyage. Like Columbus, Cabot was convinced he had reached an eastern promontory of Asia.

Still another Italian expatriate—Amerigo Vespucci—played a major role in the Spanish "Enterprise." He claimed to have made four voyages. Although two are documented and accepted, two others are questionable. Vespucci brought back significant geographical information and widely publicized his exploits throughout Europe. The historical injustice of having the New World named for him resulted from a combination of his dynamic written accounts of his voyages and his use of the term "New World" when referring to South America. European geographers named the continent *America* on the great woodcut world map of 1507. The New World has been *America* ever since.

During this period, the Portuguese continued adding to their store of knowledge about the East Indies from their travels around Africa and across the Indian Ocean. In 1499, Vasco da Gama returned from a voyage to India that established a permanent route to the Eastern lands. The following year, Pedro Alvares Cabral set out for India retracing da Gama's general route. As he crossed the Equator, sailing southwesterly, Cabral discovered the outward bulge of the Brazilian coast. This encounter explains the Portuguese heritage of modern Brazil.

Further exploration by Verrazzano, Gomez, Ayllon, Pineda, and Cartier revealed more details of North America's coastline. However, these navigators failed to find a way around or through the New World to reach the Orient. Even Magellan's success required sailing far southward to the "antarctic regions" to continue the journey west.

Because the navigators on these voyages could not determine the exact location of their discoveries, mapmakers found it difficult to construct accurate images of the new lands. Part Three of the atlas displays the rich cartographic documentation that survives from the mid-sixteenth century. These maps reveal how the newly discovered lands were gradually delineated into the two separate continents we know as North and South America.

Magellan's ship Victoria *in the Pacific on its historic circumnavigation, shown on a 1589 map by Abraham Ortelius.*

An early anonymous engraved portrait of Amerigo Vespucci.

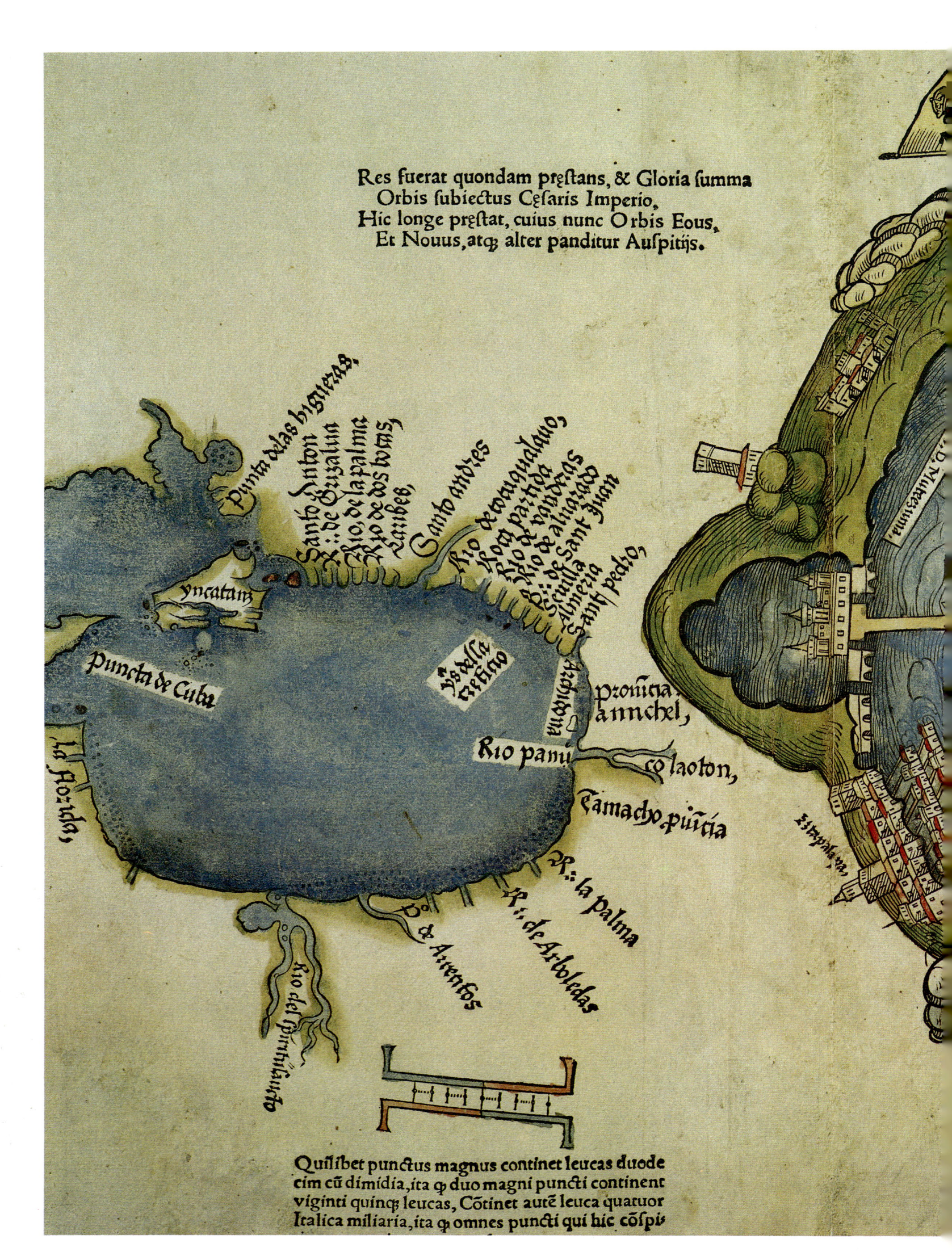

Res fuerat quondam prestans, & Gloria summa
Orbis subiectus Cesaris Imperio,
Hic longe prestat, cuius nunc Orbis Eous,
Et Nouus, atq; alter panditur Auspitijs.

Quilibet punctus magnus continet leucas duode
cim cū dimidia, ita q̊ duo magni puncti continent
viginti quinq; leucas, Cōtinet autē leuca quatuor
Italica miliaria, ita q̊ omnes puncti qui hic cōspi

PLATE 24
Hernando Cortes
(1485–1547)
Map of the Gulf of Mexico
and Plan of Mexico City
Nuremberg, 1524
Woodcut on paper,
colored by hand,
12 x 18 in. (310 x 465 mm.)
Location:
The Newberry Library,
Chicago

74

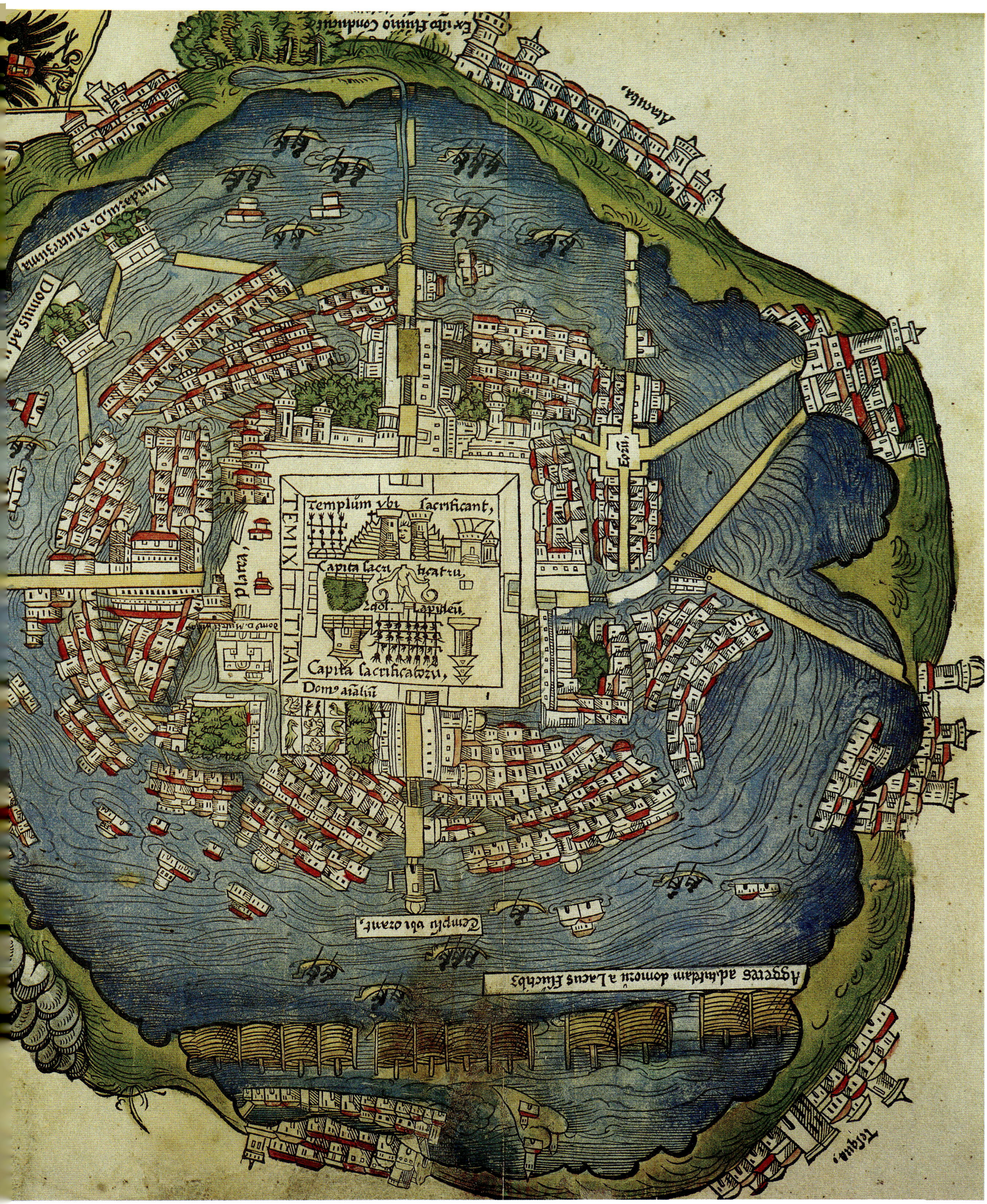
Templum vbi sacrificant,
TEMIXTITAN
platea,

HERNANDO CORTES, GULF OF MEXICO MAP & MEXICO CITY PLAN, NUREMBERG, 1524

The most successful and literate of the Spanish *conquistadores* was Hernando Cortes. His published accounts of the conquest of Mexico originated in a series of letters to King Charles V, written with the verve and piquancy for which Cortes is remembered. These two maps were issued with the Latin translation of his "Second Letter," the earliest Cortes report to have survived.

The Aztec capital of Temixtitan (Tenochtitlan on the map) in Lake Texcoco was the major indigenous urban center of North America. Cortes's city plan, first to be published of any urban site in the New World, shows the great temple of Teocalli, the two palaces of Montezuma, houses, canals, causeways, and numerous natives canoeing on the lake. The stepped-pyramid temple and plaza are surrounded by closely packed dwellings. One of the buildings along the outer shore, probably Cortes's headquarters, prominently displays a large flag with the Habsburg double eagle and crown. The Spanish captured the city in 1521, immediately razed it, and began building present-day Mexico City on the same site. Today, with the lake drained, it is one of the world's largest metropolises.

The map of the Gulf of Mexico printed on the same sheet is the first published chart of that area, the first map to use the name "Florida," and the first to show the mouth of the Mississippi River (Rio del Esprito Santo). Cortes claimed that the coastal delineation of the Mexican mainland was given to him by Montezuma; this may be true. The coastlines of Florida and what are now Alabama, Mississippi, Louisiana, and Texas are drawn from reports of the Garay expedition survivors who had escaped several hostile encounters with Indians and joined Cortes.

The leader of that ill-fated expedition was Alonso Alvarez de Pineda, while the navigator, Antón de Alaminos, was chief Spanish pilot for the Caribbean. Four caravels were outfitted by Francisco de Garay, the wealthy Governor of Jamaica. The expedition's objective was to locate a route to the South Sea (Pacific Ocean), recently discovered by Balboa in Panama. Alaminos was convinced that the west shore of the Gulf of Mexico would yield the long-sought passage to the Orient. In addition, if luck favored them, they might find gold along the way.

Pineda discovered the mouth of the Mississippi, where he spent six weeks among friendly Indians. His ships were careened inside one of the passes at the delta after a twenty-mile reconnaissance up river. After bloody skirmishes with the less hospitable natives along the Texas coast, the expedition lost a final battle to the Aztecs near today's Tampico. Pineda was killed and the survivors, a sick and starving crew on the one remaining ship, barely made Vera Cruz, from where they were brought to Cortes. Their failure to find gold or other precious metals caused later explorers to bypass the rich agricultural region from West Florida to Texas for 150 years, until the French became interested in colonization of that region.

Cortes sent these two maps to the King with his "Second Letter" detailing his exploits. They were not printed in Spain where such information was kept secret. However, this edition, published in Germany, gave the rest of Europe its first look at maps detailing these important discoveries.

References: Brown (comp.) 1952, no. 232; Church 1907, no. 53; Harrisse 1866, no. 125; 1892, 155-7, 509-10; Lowery 1912, no. 21; Sabin no. 16947; Winsor 1889, 2:364, 404; Wroth 1970, no. 34.

PLATE 24 (detail)
Hernando Cortes
Central Mexico City

JUAN VESPUCCI, WORLD MAP, ITALY, 1524

Juan Vespucci, compiler of the first double-hemispheric, polar-projection world map to be published, was the nephew and heir of Amerigo Vespucci. Juan was an expert cartographer and examiner of pilots for the *Casa de Contratación*, the governing body overseeing Spanish trade with the Indies. This novel map form was inspired by political demand and provides insight to a major international dispute of the discovery period.

When Magellan's sole surviving ship, the *Victoria*, returned to Spain in 1522, it signified the end of an era. Magellan's expedition provided the first concrete proof that the world was round. A serious dispute then arose between the kings of Spain and Portugal regarding which nation would control the Molucca Islands in the western Pacific, richest spicery of the Orient.

In 1493, when Columbus's first discoveries in the West Indies became known, the Pope was asked to mediate the conflict between Spain and Portugal over "ownership" of new overseas dominions in the East and West Indies. A Papal Bull proclaimed that lands to the west of a meridional "line of demarcation" in the western Atlantic would fall under Spain's influence, while lands east of the line would belong to Portugal. The line was revised and confirmed by treaty in 1494. It corresponds to 50 degrees west of Greenwich and crosses the mainland of South America at the mouth of the Amazon. However, the exact location of this line on the other half of the globe could not be determined because knowledge of the world's geography was limited before Magellan. Until that line was fixed, ownership of the Moluccas could not be resolved.

Charles V, King of Spain and Holy Roman Emperor, and John III, King of Portugal, eventually arranged a conference to solve the problem. Each invited knowledgeable cosmographers, cartographers, and pilots to attend. The meeting was held on the border dividing their two countries, with sessions conducted alternately at the frontier towns of Badajoz, Spain, and Elvas, Portugal. The experts brought charts and globes, and even a blank globe on which to place the East Indies and the oriental half of the line of demarcation. Juan Vespucci headed an illustrious Spanish delegation including Elcano, the surviving Magellan pilot; Sebastian Cabot; Diego Ribero; Ferdinand Columbus; and Estebán Gomez.

Vespucci's map, with its enumeration of latitude and longitude, was designed to show on a plane how the meridians from the northern hemisphere continue through the southern hemisphere, illustrating Spain's position at the Badajoz-Elvas Conference. The meridian of 315 degrees represented the treaty line. When it passed through the north pole and became the 135th degree in Asia, it bisected the Strait of Malacca between Sumatra and the Malaysian Peninsula. This clearly placed the Moluccas on the Spanish side of the line and confined the Portuguese East Indian empire to within the Indian Ocean. The hachured space along the connecting line of the 315th meridian probably refers to the alteration made between the Papal Bull of 1493 and the treaty of 1494.

The Portuguese delegation (including the cartographer Lopo Homem, who had prepared a chart of the known world), rejected the Spanish position. They presented evidence supporting their own claims, including maps, globes, and eyewitness accounts by such famous Portuguese travelers to the Far East as Duarte Barbosa. The conference deliberated extensively but was unable to settle the issue. Five years later, after the Spanish had failed twice to establish a trade route to the western Pacific via the Magellan Strait, Charles V sold his rights to King John III for what was then the enormous sum of 350,000 ducats.

As with many maps of the period, Vespucci's is a paradox of progress and retrogression. His erroneous double-peninsular representation of India contrasts with the earlier and more accurate use of many European place-names in remote locations, including Yucatan, Antigua, Guadaloupe, and Florida, which appear for the first time on a printed map.

Vespucci's work is an example of employing a new cartographic technique to solve a critical geo-political problem. While the question of ownership of the Moluccas was not settled by the map, it graphically presented Spain's position.

References: Brown (comp.) 1952, no. 62; Harrisse 1892, 533, no. 148; Lach 1965, 1:116 passim; Morison 1974, 476-77; Nordenskiöld [1897], 153, no. 40; Shirley 1983, no. 54.

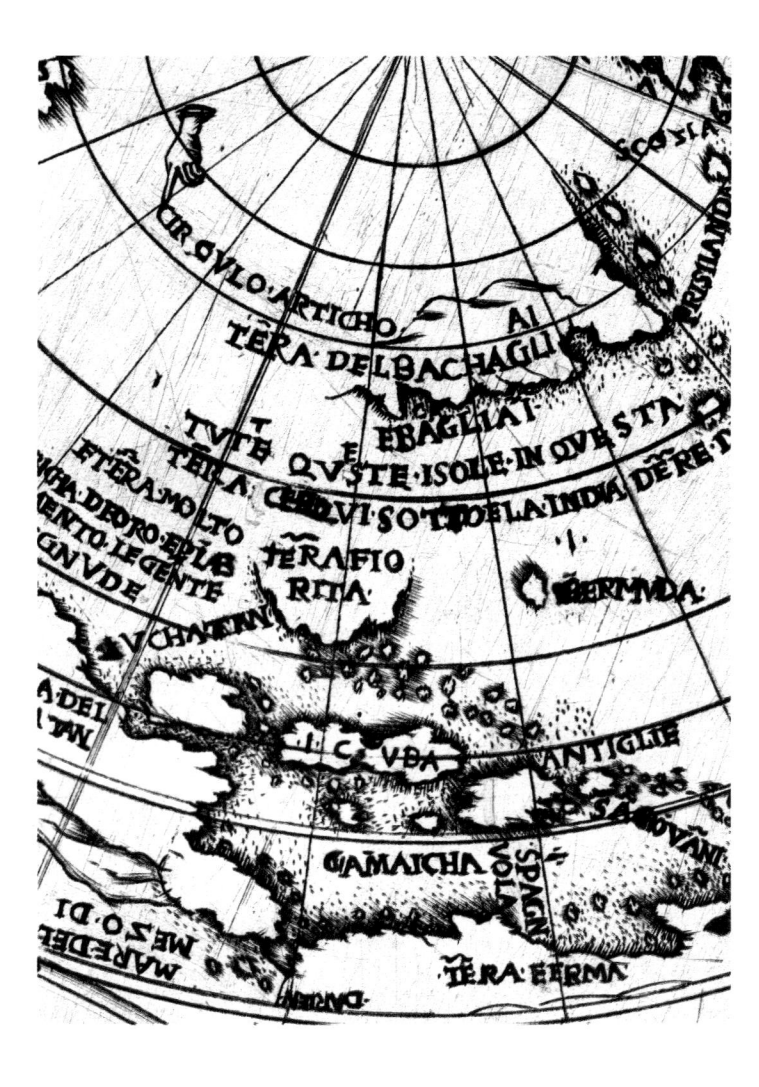

PLATE 25 (detail)
Juan Vespucci
The West Indies

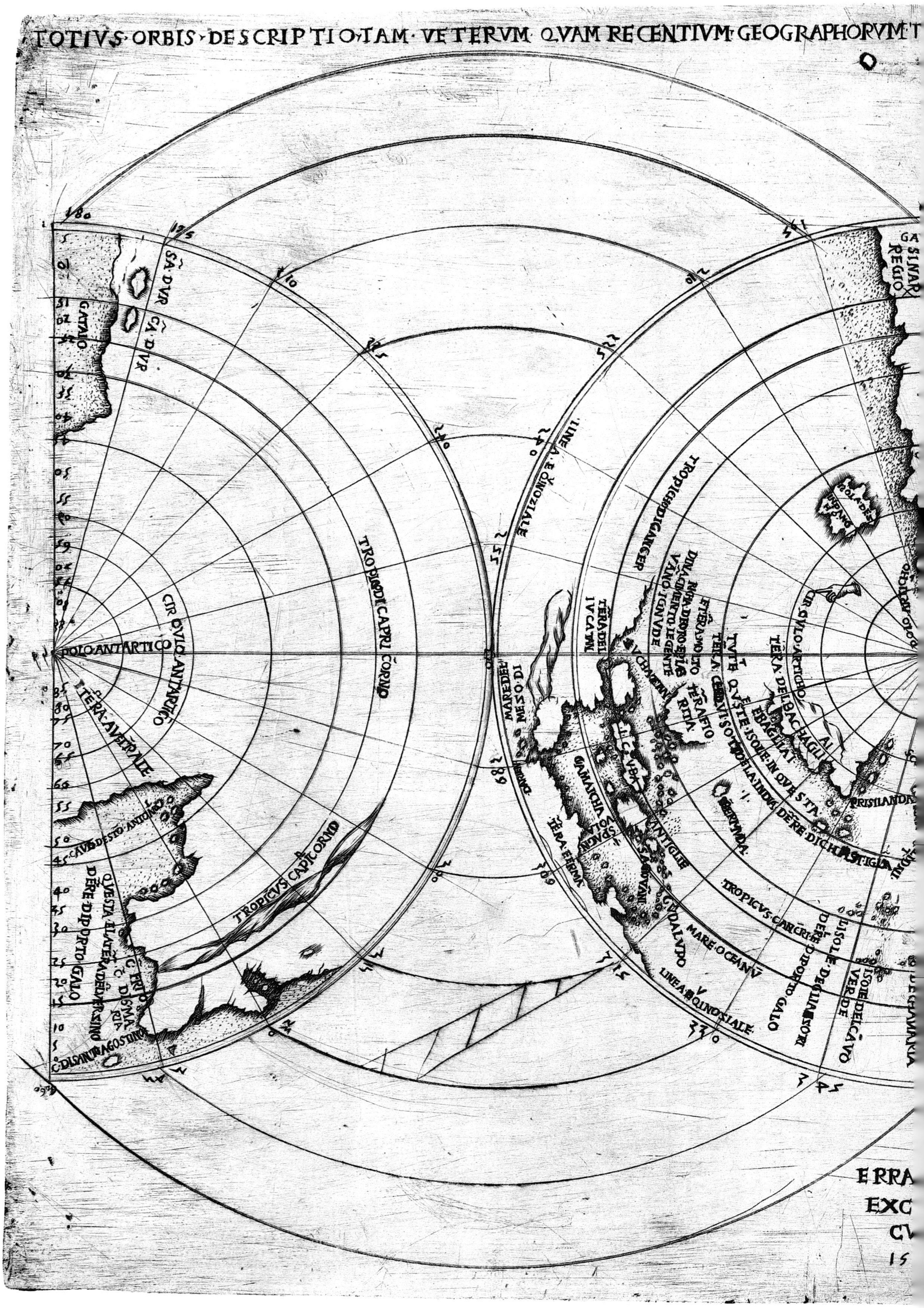

TOTIVS·ORBIS·DESCRIPTIO·TAM·VETERVM·QVAM·RECENTIVM·GEOGRAPHORVM·T

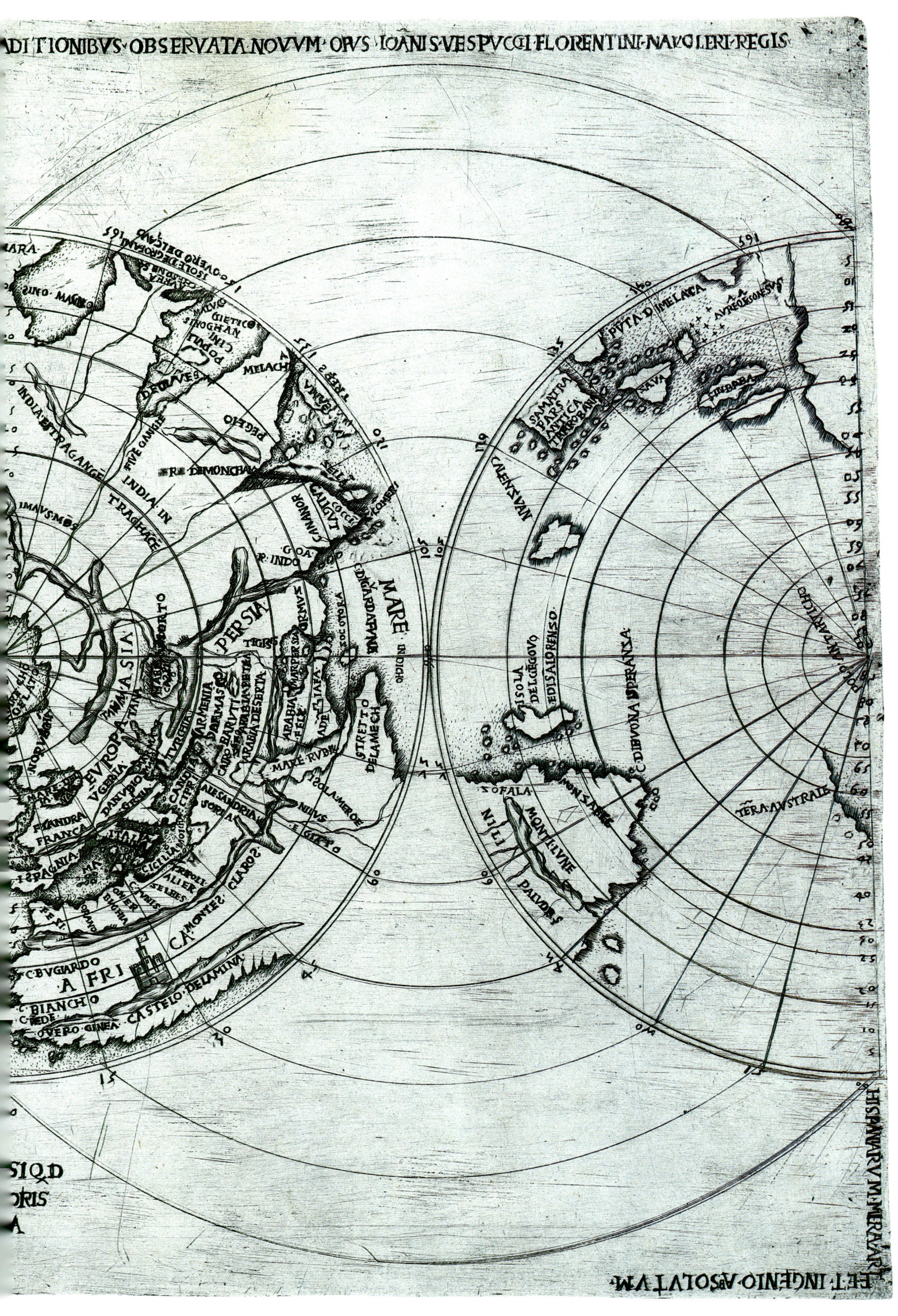

PLATE 25
*Juan Vespucci
(fl. 1512–26)*
*Totius Orbis Descriptio...
Italy, 1524*
Copperplate engraving,
15 x 11.25 in. (373 x 273 mm.)
Location:
The Houghton Library,
Harvard University,
Cambridge

ANTONIO PIGAFETTA, MAPS FROM MAGELLAN'S GREAT VOYAGE, PLACE OF ORIGIN UNKNOWN, CA. 1525

The historic first circumnavigation of the world brought its leader, Ferdinand Magellan, both premature death and a permanent place in history. These four Magellanic icons resulted from that voyage. They were among the maps accompanying a manuscript account of the three-year journey kept by Antonio Pigafetta, an Italian gentleman, adventurer, and knight of the Order of Rhodes, who sailed with Magellan. Pigafetta became one of a handful of the voyage's survivors. His passage with Magellan had been arranged by the historian Peter Martyr, a close confidant and advisor to the Spanish Court. After the harrowing voyage, Pigafetta's shipboard notes and sketch maps were superbly produced as a presentation manuscript for the head of the Order of Rhodes.

Magellan, like Columbus, Cabot, and Vespucci, made his most important discoveries after leaving his native land. As a young officer of noble birth, he had served the King of Portugal and distinguished himself in Portuguese Asia. In 1505 he fought against Muslim sea power off the African and Indian coasts, and was active through the Portuguese capture of Malacca, key port of the East Indies. After returning to Lisbon, he

PLATE 26A
*Antonio Pigafetta
(1491–1534)
The Strait of Magellan,
oriented with south at the top
Place of origin unknown, ca. 1525
Illuminated manuscript on vellum,
9 x 6 in. (225 x 150 mm.)
Location:
The Beinecke Rare Book
and Manuscript Library,
Yale University, New Haven*

80

was twice disappointed by the denial of what he believed were well-deserved promotions. The young adventurer left Portugal for Spain in 1517 to offer his services to King Charles.

Magellan's early experience in the Indies, and rumors in Lisbon of a strait opening onto the South Sea, inspired his next voyage. He presented a plan to King Charles for a Spanish expedition to reach the Spice Islands (Moluccas) by sailing west and finding a way through or around America. Charles supported the concept. Portuguese control of the route from Europe to India via the Cape of Good Hope kept that passage closed to the Spanish, making Magellan's "direct" route an attractive alternative. The route was important to the King, as the Spice Islands were on the Spanish side of the Asian extension of the Papal Line of Demarcation dividing the overseas world between Spain and Portugal. The access proposed by Magellan would allow exploitation of the rich trade in spices.

Magellan left Spain in September 1519 with five ships and 270 men. The journey was one of the most difficult in the history of exploration. Mutiny, shipwreck, and desertion quickly reduced the fleet to three ships; but the worst lay ahead. The expedition

PLATE 26B
Antonio Pigafetta (1491–1534)
Guam and native sailing vessel
Place of origin unknown, ca. 1525
Illuminated manuscript on vellum,
9 x 6 in. (225 x 150 mm.)

Location:
The Beinecke Rare Book
and Manuscript Library,
Yale University, New Haven

ANTONIO PIGAFETTA, MAPS FROM MAGELLAN'S GREAT VOYAGE, PLACE OF ORIGIN UNKNOWN, CA. 1525

remained at sea fourteen long months before finally emerging from the strait that bears Magellan's name.

The seemingly endless crossing of the Pacific that followed was horrendous. Only by eating the rats aboard ship and boiling leather sail bags did the men keep from starving. The Pacific Ocean proved to be the largest body of water on earth, not simply a short distance past the Spanish Main as Columbus and the cosmographers had thought. After sailing ninety-nine days from the Strait, Magellan and his crew were finally able to obtain fresh food at Guam. Upon arriving at the Philippines, despite earlier friendship with the natives, a skirmish between the Europeans and natives broke out and Magellan was killed. The pilot Juan Sebastian de Elcano assumed command and sailed on to the Moluccas aboard *Victoria*. He then navigated the survivors home; only one small ship and a handful of men with a profitable cargo of cloves remained of the five ships that had set sail from Spain.

Shown in Plate 26A is the earliest drawing of the "Strait of Patagonia" (Magellan). It is oriented with south at the top, a common procedure for portraying lands below the equator. To the west the great ocean is called the Pacific for the first time, reflecting this expe-

PLATE 26C
Antonio Pigafetta (1491–1534)
Mattan, the Philippines,
the site of Magellan's death
Place of origin unknown, ca. 1525
Illuminated manuscript on vellum,
9 x 6 in. (225 x 150 mm.)
Location:
The Beinecke Rare Book
and Manuscript Library,
Yale University, New Haven

dition's relief at the calmness of the sea after the turbulent South Atlantic. Plate 26B is the map of Guam, Magellan's first landfall in the western Pacific. The local natives' lateen-rigged dugout canoes, each with an outrigger, were more maneuverable than the Spanish ships and gave them a great deal of trouble. Plate 26C is the map that identifies the island of Mattan (Mactan) in the central Philippines where Magellan was killed. Plate 26D shows some of the islands of the Moluccas; the large clove tree epitomizes the objective of the entire enterprise.

Pigafetta's account is one of the most remarkable documents in the history of geographical discovery. Had it not survived, most of the knowledge it contained of the peoples of the Pacific and their way of life at the time of first contact might never have been conveyed to Europe. The maps of these newly discovered places are the only surviving graphic images from the voyage that proved beyond any doubt the world was round.

References: Lach 1965, 173-7 passim; Skelton 1958, ch. 3, ch. 7; (trans. and ed.) 1969.

PLATE 26D
Antonio Pigafetta (1491–1534)
The Moluccas with the attraction
for European exploration;
the spice tree
Place of origin unknown, ca. 1525
Illuminated manuscript on vellum,
9 x 6 in. (225 x 150 mm.)
Location:
The Beinecke Rare Book
and Manuscript Library,
Yale University, New Haven

JUAN VESPUCCI, WORLD MAP, SEVILLE, 1526

This large manuscript planisphere presents the first cartographic record of exploration in North America after the portrayal of the Cabot landfall on the La Cosa chart (Plate 10). The mapmaker, Juan Vespucci, nephew of Amerigo Vespucci, had made several voyages to American waters, according to Peter Martyr, first Spanish historian of the Indies. After Amerigo's death at Seville in 1512, Juan, who had inherited his famous uncle's maps, charts, and nautical instruments, was appointed to Amerigo's former position as official Spanish government pilot at Seville.

Juan soon became as important as his uncle in the management of geographical information from Spain's overseas activities. Juan was a member of the council to improve existing charts and the Badajoz-Elvas Commission of 1524 (Plate 25), which attempted to resolve Portuguese–Spanish claims in the East Indies. During the same year that Vespucci produced this map, he was appointed Examiner of Pilots, replacing the ubiquitous Sebastian Cabot who was then leading an expedition in Brazil.

From the headquarters they had maintained at Santo Domingo on Hispaniola since 1497, the Spanish reported little progress in the exploration of North America during the following two decades. Juan Ponce de León had been in Florida in 1513 searching for the legendary Fountain of Youth and had made an important contribution by describing the Gulf Stream. Freelance slave-raiding trips in the Bahamas and perhaps on the Florida coast were the main activities until 1520.

In that year Lucas Vasquez Ayllon, prominent Santo Domingo leader, organized an expedition to explore lands thought to exist north of Florida. He sent out a ship under Francisco Gordillo that was joined by another under Pedro de Quexos. In June 1521 they landed at the mouth of a large river, which they named after St. John the Baptist. The two captains claimed the surrounding land for Spain and, against Ayllon's orders, took 150 natives back to Santo Domingo to be sold as slaves. Quexos's ship returned safely, but Gordillo's was lost at sea. Diego Columbus headed a royal investigation that ordered the surviving Indians returned to their homeland and released.

Ayllon received a royal grant of the territory and title of governor. In 1525 he sent Quexos back to explore further, and the coastline was probed from Florida to just north of the Chesapeake. Not until July 1526 did Ayllon and his party of colonists land on the Carolina coast at the estuary they called the Jordan River. The colony was a tragic failure, but the pioneering Spaniards left their mark on the American southeastern coast via Vespucci's map. St. Helena, a sound between Beaufort and Edisto Island reflects this; it is one of the oldest place-names still in use on the Atlantic coast.

On this map Vespucci locates geographical features of which he is certain. Between Newfoundland and Nova Scotia, however, the area southward is blank until "Ayllon" (Carolina) is marked with a Spanish flag. Although none of the names from the original Gordillo and Quexos voyage appear and Ayllon's colonizing expedition had not returned when the map was produced, the second passage under Quexos in 1525 is reflected on the coastline. The area henceforth appears on maps as *Nueva Terra de Ayllon*. One interesting place-name, *Rio de sa verazanas*, is shown between the Jordan River and the Chesapeake. This implies that the explorations of Verrazzano, a Florentine sailing for the King of France, were known in Spain.

The map is considered to be either a draft or copy of the official Spanish chart kept at Seville, initially called the *padrón real* and later, *padrón general*. It was on the *padrón* that corrections and information of new discoveries were entered as reported under oath by returning pilots. Juan Vespucci provides on this outstanding manuscript map of the world the first documented details of coastal exploration north of Florida.

References: Cumming 1962, no. 7-8, 16-17, pl. 2; 1988, pl. 4; Cumming, Skelton & Quinn 1972, no. 87; Giraldi (ed.) 1954-55, no. 25; Layng (comp.) 1956, no. 105; Wroth 1970, no. 36.

PLATE 27

Juan Vespucci (fl. 1512–26)
World map, Seville, 1526

Illuminated manuscript on vellum,
33.5 x 103 in. (850 x 2620 mm.)

Location:
The Hispanic Society of America, New York

PLATE 27 (detail), *Juan Vespucci, The New World*

86

PLATE 27 (detail)
Juan Vespucci
This reproduction is the first
following the restoration
of this historic map

GEROLAMO DA VERRAZZANO, WORLD MAP, PLACE OF ORIGIN UNKNOWN, 1529

During the thirty years after Columbus's First Voyage, the Spanish, British, and Portuguese were actively exploring the New World. France, however, the fourth European maritime nation, remained uncharacteristically quiet regarding overseas exploration until 1523 when an Italian brought the French flag to North America.

In Lyon a well-entrenched enclave of Florentine bankers and merchants imported luxury goods from the Orient. They knew the traditional overland routes of European trade with Asia through the Levant could not bear the competition of Portuguese imports coming by sea around the Cape of Good Hope. Furthermore, if Spain reached the Far East by sailing directly westward to China, the Florentine merchants at Lyon would lose their trade altogether.

The solution was to search for a sea passage that allowed its discoverer to control trade with the Orient. A group of powerful and wealthy Florentine *Lyonnaise* agreed to sponsor a voyage westward in search of a route to Cathay. They engaged Giovanni da Verrazzano, a Florentine pilot, to lead the enterprise, but needed the authorization of the King of France. François I, who had been occupied in regional warfare and in developing a brilliant court, found his treasury in need of immediate funds. He promptly approved plans for the proposed westward voyage. Francois had no inhibitions over poaching on the Portuguese or Spanish preserve; in fact, he was incensed at the arrangement between those two nations and at the Pope who had divided the world between them.

In January 1524, Verrazzano and his fifty-man crew set off in their 100-ton caravel. They returned to Dieppe six months later after an extraordinary cruise that changed the maps of America and affected the direction of exploration for decades to come. Their first landfall was near today's border of North and South Carolina. As this area appeared impenetrable, they proceeded northeasterly in their search for the passage through to the Orient. After passing Cape Fear and Cape Lookout, Verrazzano sailed the long reaches outside the Outer Banks of North Carolina. He sighted Cape Hatteras and continued northward, traveling the 150 miles around Pamlico and Albemarle Sounds. The large body of water inside the Outer Banks was visible from the ship but no inlet was found.

During this run Verrazzano reached the incredible conclusion that Pamlico Sound was a great sea connecting directly with the ocean whose western shore was Cathay! The most dramatic feature of the Verrazzano map is this vast nonexistent protuberance of the Pacific Ocean that appeared as the "Sea of Verrazzano" on maps and globes for over fifty years.

Wishful thinking no doubt impelled Verrazzano to place this enormous waterway in the middle of North America, contained at the east by the narrow isthmus of the Carolina Outer Banks. It must have been the same motivation that caused European mapmakers to accept this delineation for many years. Richard Hakluyt, Elizabethan propagandist and promoter of British overseas enterprise, included a map based on Verrazzano's in his influential *Divers Voyages Touching the Discoverie of America* published in 1582. Sir Walter Raleigh's colony at Roanoke was established on an island in Pamlico Sound presumably because it would be a strategic location on the main route to Cathay when the passage was discovered.

The Verrazzano brothers, Giovanni the explorer and Gerolamo the mapmaker, combined to create an image that persisted to the end of the century. Unfortunately, they never found the passage into the sea that bore their name. On Giovanni's third voyage, while searching for a strait through Panama to the Southern Ocean, a ghastly final scene was played out. The explorer went ashore on an unnamed Caribbean Island with a small party and was captured by a band of hostile Indians within sight of his horrified but helpless brother still aboard ship. Giovanni Verrazzano was said to have been killed, butchered, and eaten. Previous encounters with cruel Spanish conquerors had obviously affected the Indians' protocols of hospitality.

References: Cumming 1962, 9-10; 1988, pl. 5; Ganong 1964, 99-133; Harrisse 1892, no. 165; Lowery 1912, no. 30; Stevenson 1903, no. 12; Stokes 1915-28, 2:pl. 13; Winship 1900, no. 229; Wroth 1970.

PLATE 28 (detail)
Gerolamo da Verrazzano
The New World and the Sea
of Verrazzano

89

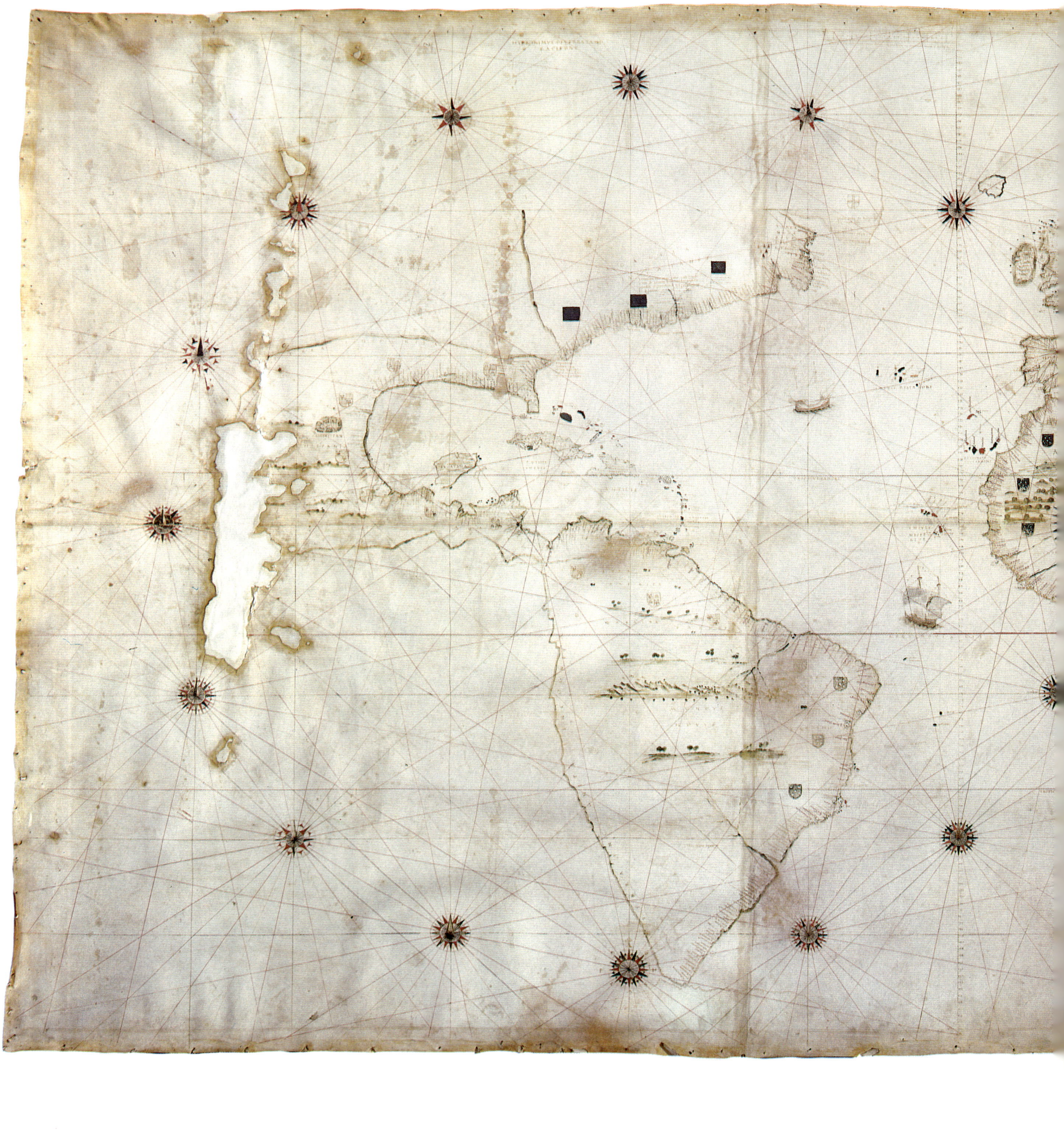

PLATE 28
Gerolamo da Verrazzano (fl. 1522–29)
World map
Place of origin unknown, 1529

Illuminated manuscript on vellum, 51 x 102 in. (1275 x 2550 mm.)
Location:
Biblioteca Apostolica Vaticana, Vatican City

DIEGO RIBERO, WORLD MAP, SEVILLE, 1529

Diego Ribero's map is justifiably considered the finest cartographical production of its age. The mapmaker, whose name in its Portuguese form is Diogo Ribeiro, was a Lusitanian at Seville in the service of Charles V of Spain. He succeeded Sebastian Cabot as Pilot Major, a post for which he was highly qualified, having navigated to India for both Vasco da Gama and Albuquerque.

The map is a copy of the official Spanish *padrón general*, the most accurate, continually updated delineation of the known world. It was Ribero's responsibility to examine all returning Spanish pilots and to incorporate the latest discoveries, corrections, and revisions on the master map. For example, it was Ribero who first included important new information from the survivors of Magellan's circumnavigation.

In North America coastal details are shown from reports of the unfortunate attempt by Lucas Vasquez de Ayllon in 1526 to colonize the Carolina coast. More than 500 prospective settlers had set out with Ayllon, including twenty black slaves, the first known to have been brought to the continent. Through a variety of misfortunes, which included Ayllon's death, bitter quarrels among his successors, and the neighboring Indians' revenge for their mistreatment, the colony failed. Ribero's southeastern coastline of North America includes contours and place-names reported by the 100 survivors who made their way back to Santo Domingo.

Estebān Gomez, another experienced Portuguese pilot in the service of the Spanish King, had commanded the ship of Magellan's squadron that deserted the circumnavigator at the strait which bears his name. Although initially imprisoned for such treachery, Gomez was released and eventually sent back to sea by the King. In September 1524, he embarked for North America with one small ship to search for the elusive passage to China still thought to exist somewhere between Florida and the Cape Breton-Newfoundland area. Verrazzano (Plate 28) had returned to France in July of the same year after failing to find the passage. Gomez reconnoitered the North American coast from Nova Scotia to Nantucket with no better luck than his predecessor and eventually returned to his port of Coruna. Ribero incorporated into the *padrón* Gomez's information on the area from the mid-Atlantic coast to Cape Breton.

The place-names from the Ayllon and Gomez voyages, together with the revised northeasterly curving coastline of North America, remained on Spanish maps for over a century. Noticeably omitted is any reference to Verrazzano's explorations. Political expedience precluded the official Spanish map mentioning a French voyage of reconnaissance in Spanish territory.

The map is rendered with considerable artistic accomplishment. Finely executed trees and animals, including many exotic species new to Europeans, and beautiful ships and compass roses complement the su-

PLATE 29
Diego Ribero (d. 1533)
World map
Seville, 1529
Illuminated manuscript on vellum,
33 x 80 in. (850 x 2050 mm.)
Location:
Biblioteca Apostolica Vaticana, Vatican City

perb cartography. The combination of art and science is underscored by detailed drawings of the astrolabe and quadrant at bottom right and left, and the elaborate declination scale west of America. It was the instrument maker and experienced navigator, Diego Ribero, who first used such scientific decoration on his maps, replacing the traditional religious themes of earlier mapmakers.

In this replication of the royal map of Spain, the results of the first phase of the Age of Discovery are clearly depicted. Here is the world as "re-made" by the exploits of Columbus, Cabot, Corte-Real, Vespucci, Balboa, Magellan, and other courageous leaders who searched until they found many of the secrets hidden from earlier Europeans.

References: Cumming 1962, 8; 1988, 11-17; Cumming, Skelton & Quinn 1972, 104; Ganong 1964, 149-52; George 1969, 62-64; Harrisse 1892, 573-75; Lowery 1912, no. 31; Mollat du Jourdin 1984, no. 37; Wroth 1944, no. 40.

PLATE 29 (detail)
Diego Ribero
*Florida, the West Indies,
and the Spanish Main*

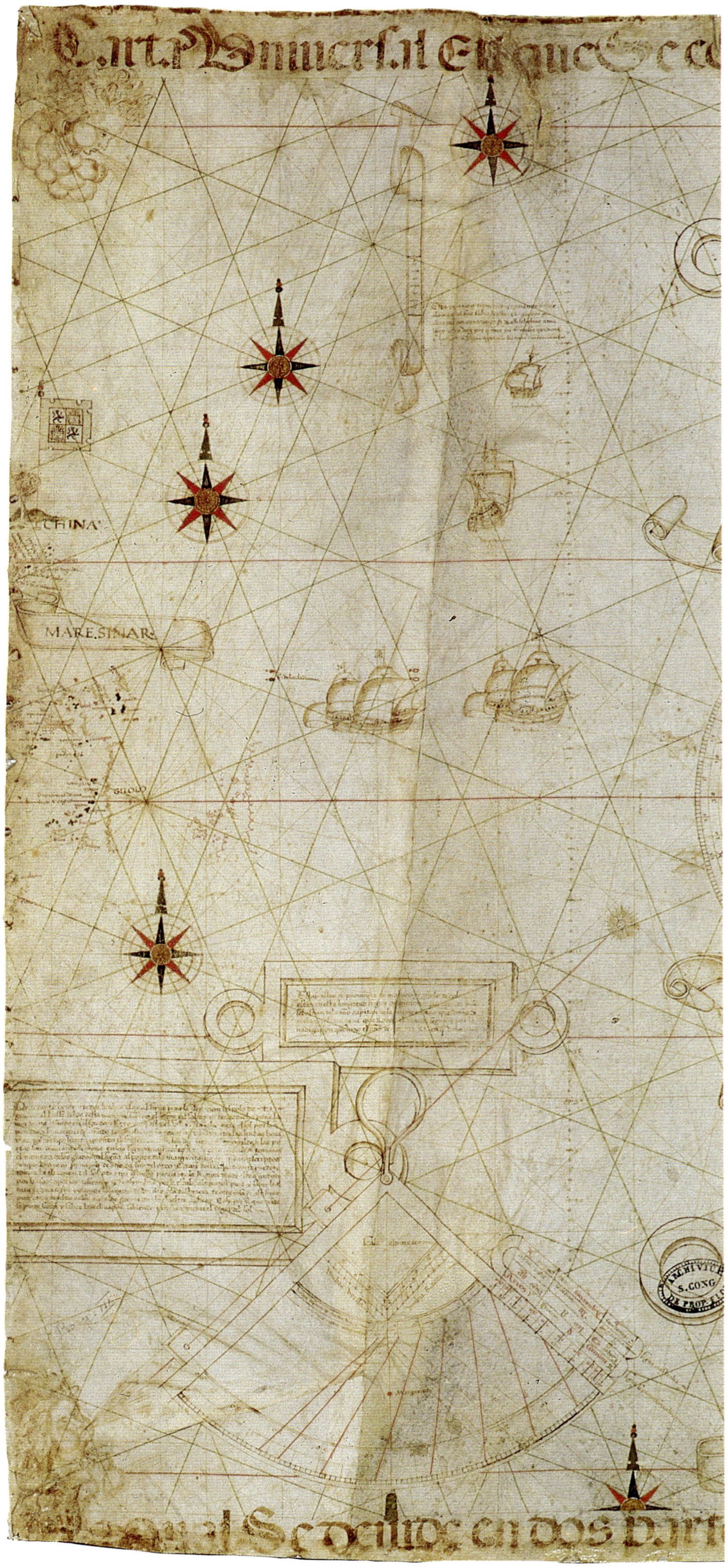

PLATE 29
Diego Ribero
World Map
*(extreme left portion,
remainder on following pages)*

PLATE 29
Diego Ribero
World map
(extreme left portion on previous page)

95

DIEGO RIBERO & GIOVANNI BATTISTA RAMUSIO, THE NEW WORLD, VENICE, 1534

Only three examples of this remarkable woodcut are known to have survived, although many more copies exist of the book it accompanied. It is the only printed map directly derived from the official Spanish master chart kept in Seville and the most accurate delineation to have appeared for public use. This and the *West Indies* map published by Peter Martyr in 1511 (Plate 19) are the sole published maps of America originating from the Seville school of cartography during the early exploration period.

Although it is a simple composition when compared with the elaborate manuscripts of the time, the delineation is nevertheless remarkably well presented. At the top the work continues the tradition of naming Greenland "Labrador," after João Fernandez, the farmer (or *labrador*) from the Azores who traveled there in 1501. Today's Newfoundland–Labrador area is called "Bacalaos" (Portuguese for codfish). The area's rich fishing banks had been the objective of European voyages since the late fifteenth century.

The descending coastline of North America follows the contours reported by Estebãn Gomez upon returning from his reconnaissance voyage of 1525. His name appears on the New England coast he explored, just as "Licenciado Ayllon" is shown near the Carolina shore where the Santo Domingo governor tried to establish a colony. The name of Florida is in place and the shape of the peninsula is strikingly accurate. There is a careful rendering of the north coast of the Gulf of Mexico, indicating the unnamed mouth of the Mississippi, which reflects Alonso de Pineda's voyage. The West Indies are improved, but Yucatan appears as an island, an error that persisted for years.

Southward the Central American coastline clearly reveals a narrow isthmus and indicates *Mare del Sur* (the Pacific Ocean), based on Balboa's discovery. Panama is named on the Pacific side, with a caravel depicted standing off shore; on the Caribbean coast the Panamanian ports Nombre de Dios and Darien are also shown.

The eastern part of South America is called "Brasil" and the west, "New Castile or Peru." The east coast configuration is clearly drawn, indicating the estuaries of the Orinoco, the Amazon, and the "Jordan" (Rio de la Plata) rivers. At the south end of the continent, the Strait of Magellan is drawn and named for the first time on a printed map. Although a world map issued two years earlier by the French cartographer Oronce Fine indicated a passage at the bottom of South America, this map is the first to identify and clearly depict the historic strait. Magellan's epoch-making voyage, completed posthumously in 1521, provided the first tangible proof that global theories pursued by Columbus and others were correct, even if their distances were not.

Because Spain's policy prohibited publishing geographical information, the authorities did not sanction the printing of this map. It was produced in Italy by Giambattista Ramusio to accompany an edition of the early histories of America by Peter Martyr and Oviedo. Since it does not carry the maker's name, it has been known either as the "Ramusio" or the *Sumario* map, after the editor or the title of the book. Although a modest production, this woodcut version of the American portion of Ribero's planisphere is an historic monument in cartography. It brought important new information to a broad audience who had no access to the lavishly produced manuscript maps available only to royalty, nobility, merchants, or bankers.

References: Nordenskiöld [1889] 1961, 106-07; Stokes 1915-28, 2:26, pl. 7; Winsor 1889, 2:223; Wroth 1970, no. 59.

PLATE 30
*Diego Ribero (d. 1533)
& Giovanni Battista Ramusio
(1485–1557)*
Terre Firma
& the West Indies
Venice, 1534
*Woodcut on paper, colored by hand,
21 x 17 in. (530 x 425 mm.)
Location:
Arthur Holzheimer collection*

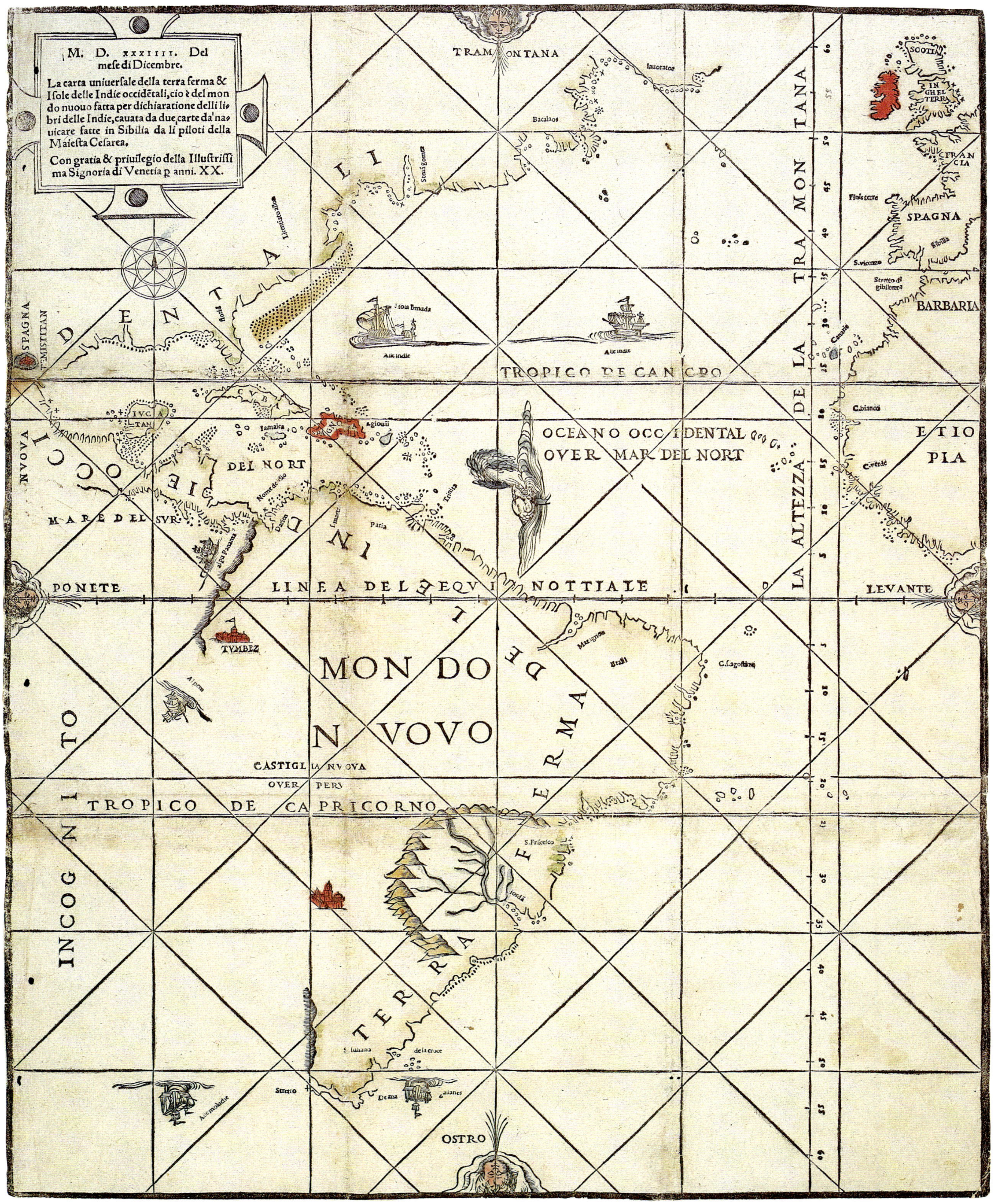

M. D. XXXIIII. Del
mese di Dicembre.

La carta uniuersale della terra ferma &
Isole delle Indie occidetali, cio è del mon
do nuouo fatta per dichiaratione delli li
bri delle Indie, cauata da due carte da'na
uicare fatte in Sibilia da li piloti della
Maiesta Cesarea.

Con gratia & priuilegio della Illustrissi
ma Signoria di Venetia p anni. XX.

TRAMONTANA

OCCIDENTALI

SPAGNA

TROPICO DE CANCRO

NVOVA

INDIE OCCI

MARE DEL SVR

DEL NORT

OCEANO OCCIDENTAL
OVER MAR DEL NORT

LA ALTEZZA DE LA TRAMONTANA

SCOTIA
IN GHEL
TERRA

FRANCIA

SPAGNA

BARBARIA

ETIO
PIA

PONITE

LINEA DEL EQUINOTTIALE

LEVANTE

TVMBEZ

MONDO
NVOVO

TERRA FERMA DEL

INCOGNITO

CASTIGLIA NVOVA
OVER PERV

TROPICO DE CAPRICORNO

OSTRO

SEBASTIAN MUNSTER, THE NEW ISLANDS, BASEL, 1546

Münster was professor of Hebrew at the University of Basel and a prominent German mathematician, cartographer, and cosmographer. Among his publications were an edition of Ptolemy's *Geography* in 1540, followed in 1544 by his *Cosmography*, the influential first German description of the world. Both works were illustrated by a series of woodcut maps that included this delineation. It was the first printed map devoted to the Western Hemisphere. Its innovative continuous coastline through North, Central, and South America emphasized a definite separation of the New World from Asia. The publications were so popular that together they appeared in forty editions, making them best sellers of the sixteenth century. As a result, this woodcut of America was more widely circulated than any map of the New World of the time.

An important characteristic of Münster's map is the enormous Sea of Verrazzano and northeasterly trend of North America. This resulted originally from Verrazzano's exploration of the east coast, when he convinced himself that he was seeing the South Sea (Pacific) just west of the Carolina Banks (Plate 28). That misconception, which this map helped to perpetuate, encouraged the French and later the British to believe that a passage could be found through North America to China.

At the same time Münster kept alive many earlier traditions. While not actually locating the papal line of demarcation between Spain and Portugal's overseas domains, the flags of those two nations appear on their respective sides of where the line would cross Brazil. The language differences in South America today originated with this deployment. Columbus's reports of "abundant gold and pearls at Paria" on South America's north coast and the Vespuccian reference to cannibals in Brazil are retold. Magellan's first circumnavigation of the globe is celebrated by a depiction of his ship *Victoria*, the only survivor of his five-ship fleet. Münster also labeled the strait with the name of its discoverer and followed Magellan in calling Patagonia, "Land of Giants." Even medieval Marco Polo is remembered; off the China coast is a note on Polo's "archipelago of 7448 islands," and a depiction of Cipangu (Japan), strikingly reminiscent of Pizzigano's *Antilia* (Plate 3), of over a century earlier.

In showing North and South America as continents separate from the Old World, Münster improved upon previously published maps. On the other hand, such distinguished cartographers as Giacomo Gastaldi and the Italian School continued to link America to Asia for another twenty-five years. This charming woodcut with its geographic advances and retrogressions is one of the best-known images from the period.

References: Bagrow/Karrow 1990, no. 58:117; Lowery 1912, no. 46; McCorkle 1985, no. 6; Nordenskiöld [1889] 1961, fig. 73; Schwartz & Ehrenberg 1980, 45; Wroth 1944, no. 46; Wroth 1970, No. 47.

PLATE 31
Sebastian Münster (1489–1552)
The New Islands
Basel (Switzerland), 1546
Woodcut on paper, colored by hand,
10 x 14 in. (257 x 348 mm.)

Location:
James Ford Bell Library,
University of Minnesota,
Minneapolis

BATTISTA AGNESE, WORLD MAP, VENICE, 1542

Battista Agnese's highly artistic work has been admired for over four centuries. Formerly, it was believed that the small format of his charts rendered them less important than the large planispheres of the period. However, Agnese's carefully drawn diminutive maps conveyed significant new information of geographical discoveries and, in some instances, were the first to do so.

Sixty-five elegantly designed and executed Agnese atlases have survived, indicating he was one of the most prolific mapmakers of the sixteenth century. During a thirty-year career he is reputed to have produced over one hundred atlases averaging ten charts in each. Unfortunately, however, we know nothing of the man beyond what is found on his maps. A Genoese working in Venice from 1535–65, Agnese signed and dated a few of the charts, making his work identifiable; however, scholars have long searched in vain for further details of his life.

The Agnese atlases began with a world map on a plane projection made up of three sheets with considerable overlapping, such as the one shown here. One sheet focused on the Pacific region, another on the Atlantic, and a third on lands bordering the Indian Ocean. There followed a "normal portolan" (the Mediterranean area), augmented by the Black Sea and the British Isles in from four to six maps. The contents were completed with an oval projection world (Plate 33) and occasionally an Atlantic hemisphere focusing on America, Europe, and Africa.

Agnese's charts are most significant in the mapping of America and Asia; the coasts of Europe and Africa had been known before his time. The Sevillian cartography of Diego Ribero (Plate 29) was his principal source, and Agnese's three-sheet world map covered the same geographical area as Ribero's. The North American coastline, while not showing results of Cartier's voyage, reflects the Spanish expeditions of Ayllon and Gomez on the Atlantic seaboard and those of Cortes and Pineda in the Gulf of Mexico. The voyages of Niño and Avila on both Central American coasts are included, and on the Pacific side of South America, information is derived from Pizarro's discoveries.

The timeliness of Agnese's cartography is seen in the peninsula of California. He is the first European mapmaker to include the news of Francisco de Ulloa's Cortez-sponsored voyage northward from Acapulco to the Gulf of California (Sea of Cortez). Ulloa named the Gulf the "Vermilion Sea" and discovered the mouth of the Colorado River during this reconnaissance in 1539–40, two years before Agnese's atlas was completed.

On his Pacific sheet, Agnese became the first chartmaker to indicate water depth, although it is not rendered in the form of soundings as we know them today. The unit of measurement was the *braza*, (the length of two arms outstretched). This term, used since earliest times, referred to the amount of line hauled in by the leadsman to determine the depth of water when a sounding was taken. The inscription on the map at the head of the Gulf reads, "Vermilion Sea in which there are eleven *brazas* in the channel at high water and more than eight at low water."

Agnese, whose maps were seen by more people than those of any other manuscript chartmaker of the century, played a key role in disseminating news of

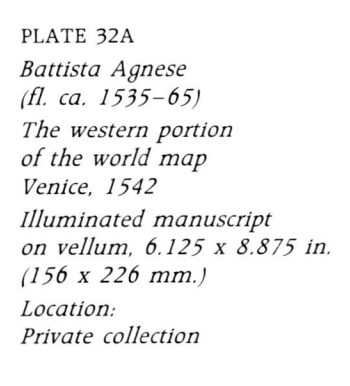

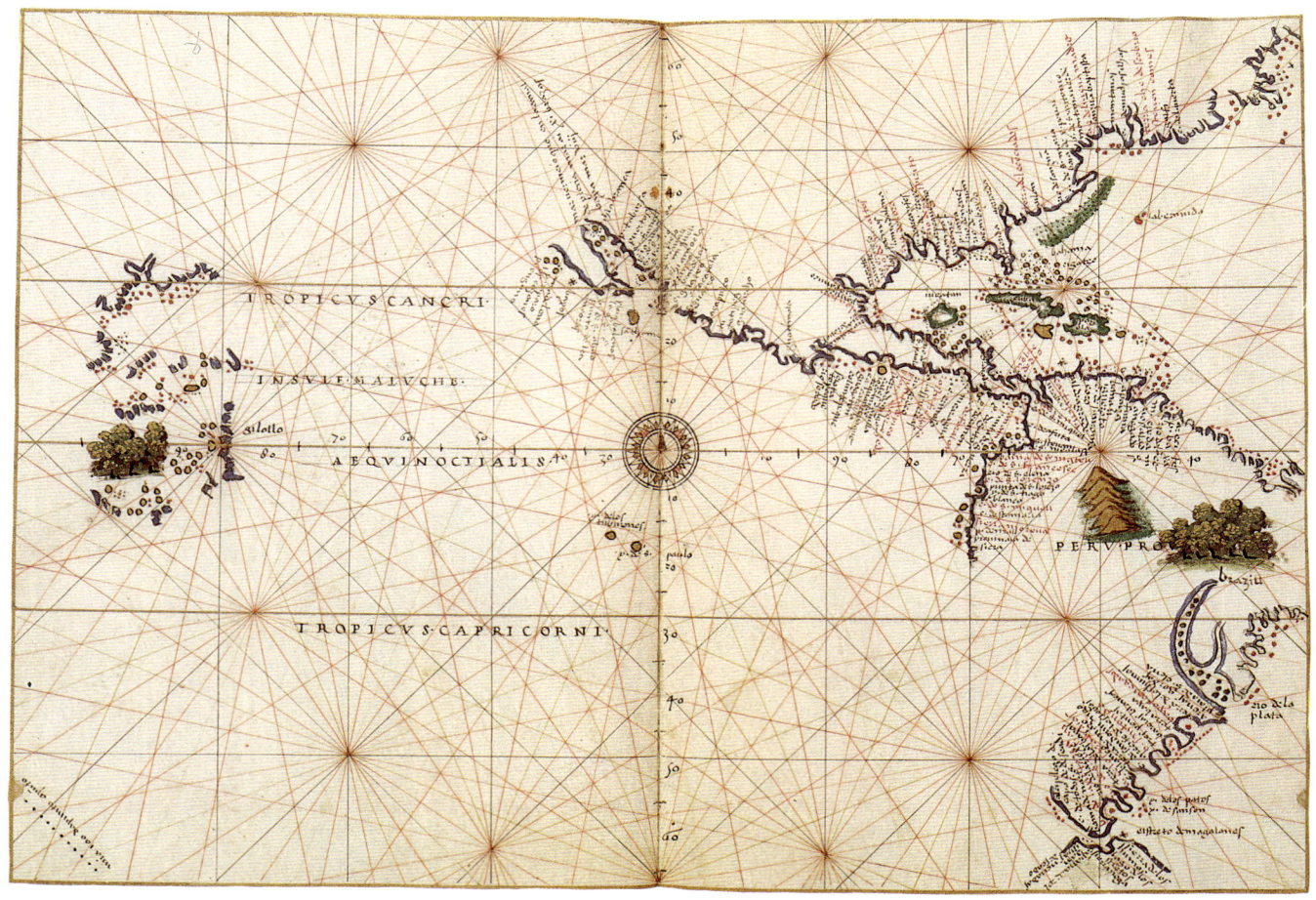

PLATE 32A
Battista Agnese
(fl. ca. 1535–65)
The western portion
of the world map
Venice, 1542
Illuminated manuscript
on vellum, 6.125 x 8.875 in.
(156 x 226 mm.)
Location:
Private collection

overseas discovery with his exquisite work. They remain a permanent record of the moment when Europe was completing its knowledge of how the earth's continents and oceans were arranged.

References: Brown (comp.) 1952, 90; Ganong 1964, 163-64; Harrisse 1892, 236; Howse & Sanderson 1973, 32-33; Layng (comp.) 1956, 225-30; Lowery 1912, no. 40; Skelton 1958, 46; Wagner 1931, no. 24; 1937, 4; Wroth 1970, no. 46.

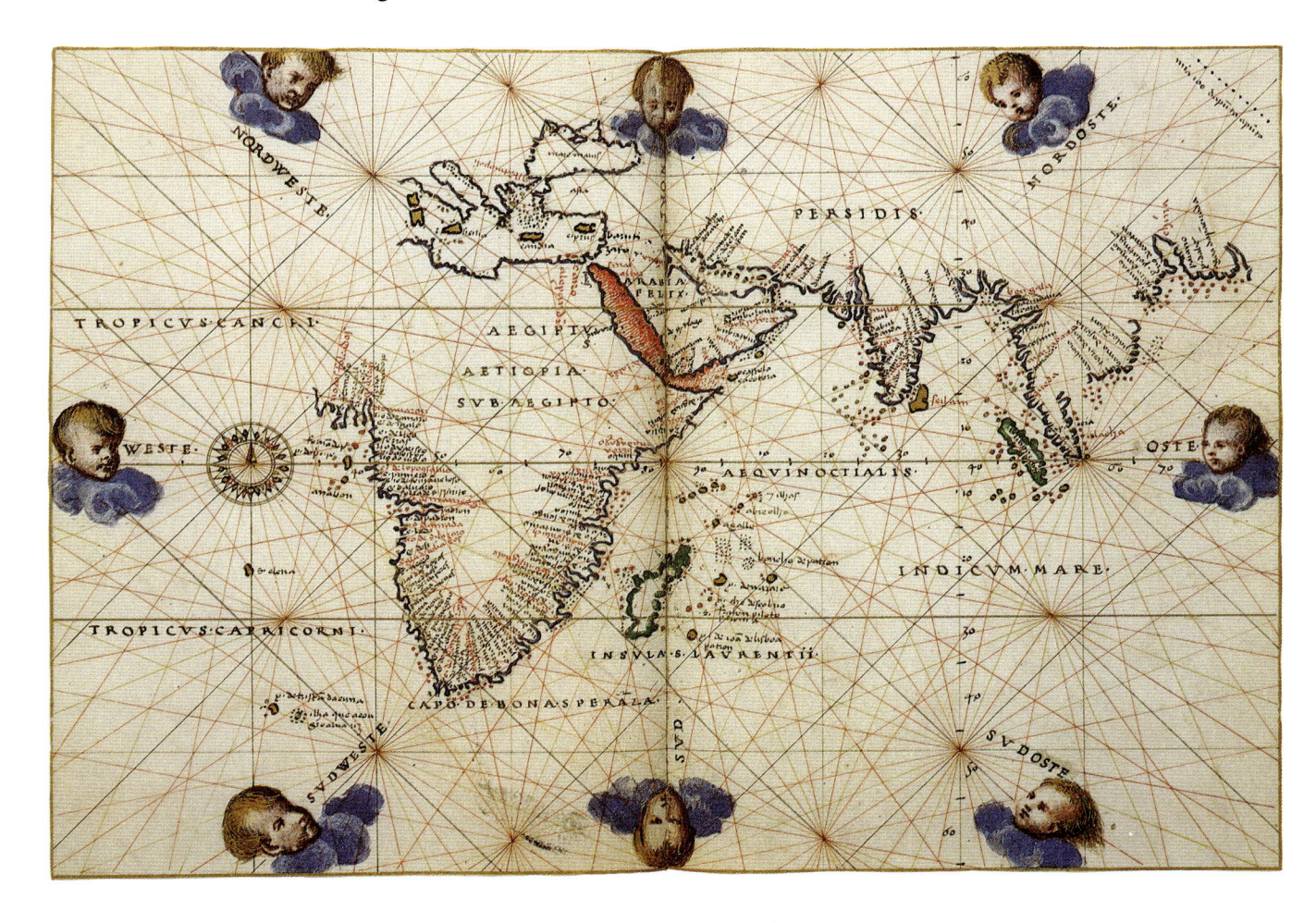

PLATE 32B
*Battista Agnese
(fl. ca. 1535–65)
The eastern portion
of the world map
Venice, 1542
Illuminated manuscript
on vellum,
6.125 x 8.875 in.
(156 x 226 mm.)
Location:
Private collection*

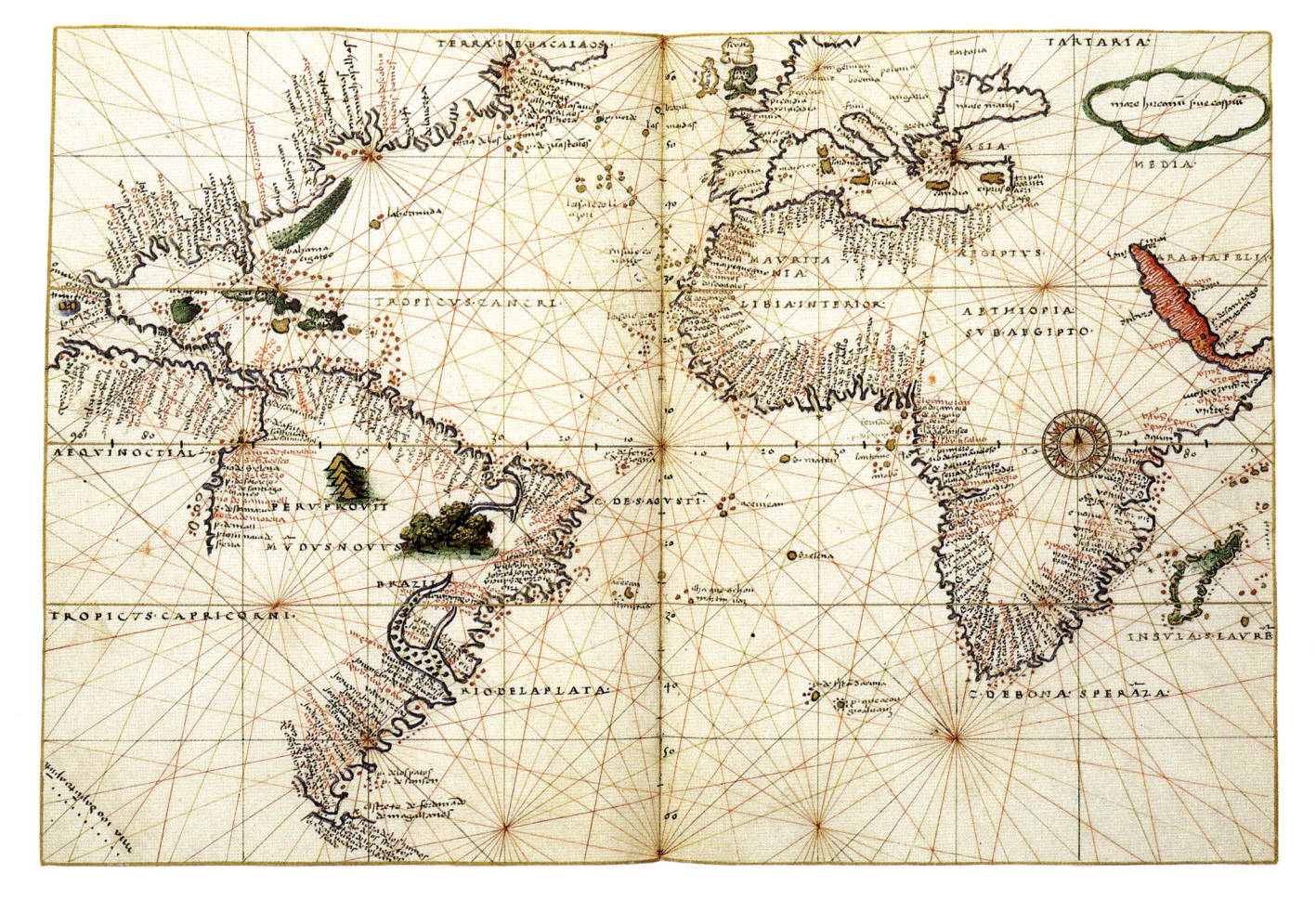

PLATE 32C
*Battista Agnese
(fl. ca. 1535–65)
The central portion
of the world map
Venice, 1542
Illuminated manuscript
on vellum, 6.125 x 8.875 in.
(156 x 226 mm.)
Location:
Private collection*

BATTISTA AGNESE, OVAL WORLD MAP, VENICE, 1542

The most distinctive of Agnese's works is this world map, which some historians have erroneously considered merely derivative. Constructed on an oval projection introduced by Rosselli (Plate 17), it is clearly in conflict with his three-sheet plane projection chart (Plate 32). Agnese presented the Spanish version derived from Ribero (Plate 29) on the three-sheet chart, and the French Verrazzanian concept (Plate 28) on this oval projection. These opposing theories were both popular at that time.

The oval world map is distinguished by the silver and gold lines crossing the oceans. Silver represents the routes from Spain to the Moluccas via the Strait of Magellan and homeward around the Cape of Good Hope. Gold depicts the track of the Spanish fleet to Panama and Peru. Agnese was the earliest mapmaker to use this device, which became his trademark.

While the classical world map of Ptolemy formed its base, many refinements in this map reflect progress in geographical knowledge. The parts of the world that had been traveled by Europeans show detailed configurations, while the unexplored areas are enclosed by featureless coastlines. America is clearly separated from Asia and, although the Indian Ocean mapping has not been perfected, the huge austral continent of Ptolemy has disappeared. There is also a *Terra Nove* extending north of Scandinavia to the North Pole that is unique to Agnese.

The southern New England coast from New York to Cape Cod, lost on the plane projection chart that follows Gomez and Ribero, appears here delineated from Verrazzano's description. Thus we have a continuous coastline from Florida to the ill-defined "Codfish Country" of Labrador and Newfoundland. Agnese continued to believe in the possibility that the Sea of Verrazzano, still featured here, was based on fact. If so, an Atlantic entrance might be found to unlock a short route from Europe to China. Ten years later he continued to include this false sea that also remained on other European maps for another generation.

Displaying as it does the peninsula and Gulf of California, Agnese's oval map could not have been completed much earlier than 1542. That information, and the discovery of the mouth of the Colorado River, a representation of which is shown, came from the Ulloa voyage of 1539-40. It is unlikely that the news could have reached Venice from the Pacific within one year.

The two Agnese world maps provide us with the same unique juxtaposition of two conflicting delineations that his contemporary clients contemplated, drawn during the same year by the same cartographer. Perhaps this combination, together with the beauty of Agnese's draftsmanship and decoration, account for his being the most popular manuscript chartmaker of the era.

References: Fite & Freeman 1926, frontis., no. 17; also, see pl. 32.

PLATE 33
Battista Agnese (fl. ca. 1535–65)
Oval world map
Venice, 1542

Illuminated manuscript on vellum, 6 x 9 in. (156 x 226 mm.)
Location:
Private collection

SEBASTIAN CABOT, WORLD MAP, ANTWERP, 1544

This cartographic monument, produced for commercial publishers in Seville and engraved at Antwerp, was compiled by Sebastian Cabot, one of the most celebrated names in the Age of Discovery. Its purpose was to bring before the public in printed form a large informative world map based on the official Spanish original, such as Diego Ribero's manuscript (Plate 29).

Previous printed maps of the world were on a small scale, often accompanying a new edition of a classical geography or collection of voyages. The notable exception, by Waldseemüller (Plate 16), had been published thirty-seven years earlier. Cabot's map became a classic, both because of his authorship and the spectacular iconography that complements the delineation.

His active career, spanning six decades as an explorer and advisor to kings, began when he accompanied his illustrious father John on the voyage from Bristol to the mouth of the St. Lawrence in 1497. While not perfectly documented, that voyage is credited with the first discovery of North America by a European since the Norsemen five centuries earlier. Because some of the younger Cabot's claims are difficult to verify, and with his flair for self-aggrandizement, he became in his day as controversial as he was famous.

In 1509 Sebastian again sailed for the English seeking a northwest passage to the Orient. Later he was to claim, perhaps justifiably, that he discovered Hudson's Bay (a century before Hudson). By 1512 Cabot was made cartographer to England's Henry VIII. Soon afterwards, however, King Charles V of Spain retained him to be a member of the Spanish Council of the Indies, Pilot Major, and examiner of pilots at Seville.

Cabot commanded a three-ship Spanish expedition to the Moluccas in 1524, which he diverted to the Rio de la Plata region of South America to hunt for gold. After three years and no gold, he returned to Spain in disgrace and was banished. Yet by 1527 this remarkable explorer was back as Pilot Major. At age 70 Cabot was reappointed naval advisor to Edward VI of England. At the time of his death a decade later, he was organizing yet another voyage to find the northwest passage.

Sebastian Cabot indicates the site of his father's landfall in America at 50 degrees north latitude, while mapmakers, including La Cosa and Ribero (Plates 10 and 29), had placed it 10 degrees farther north. It has been speculated that this was an intentional maneuver to support English claims to eastern Canada and to oppose the French. At the same time, by including the St. Lawrence, Cabot reflects Cartier's presence there in 1534-35.

In the Southwest, Mexican-based explorations of North America by Francisco de Ulloa in 1539 in the Gulf of California, and Coronado in 1540–42, also contributed to the delineation. On the map, lower California is shown as a peninsula, and the mouth of the Colorado River appears. In the Pacific, legends summarize contemporary knowledge of Mexico and Peru. Cabot's emphasis in South America is on the Amazon and the Rio de la Plata, both of which he had personally explored.

Strangely enough, the map's weakest point is its failure to show Europe accurately. From the Mediterranean to Scandinavia, including the British Isles, the configurations are incorrect. Overall, Cabot's attempt to develop an innovative projection was notably unsuccessful. Nonetheless, viewers have always been attracted by the splendid illustrations of fauna, both actual and imaginary, and of fabulous people accompanied by legends from early medieval sources. The Asian lore includes a large vignette of the great Khan along with a lengthy description of China and Japan derived from Marco Polo.

Sebastian Cabot's voyages and his important service to English and Spanish kings as cosmographer and mapmaker during his long life made him a towering figure of the discovery era. This planisphere is his only surviving map. It has long been known only from this copy, although recently another example was discovered in Weimar, East Germany.

References: Bagrow/Karrow 1990, no. 17:6.1; Fite & Freeman 1926, no. 18; Ganong 1964, 230ff; George 1969, 91; Kelsey 1987, 41ff; Layng (comp.) 1956, 259; Lowery 1912, no. 44; Schilder 1986-88, 2:23-26; Shirley 1983, 81, pl. 69; Winship 1900, 39; Winsor 1889, 3:22; Wroth 1970, no. 64.

PLATE 34 (detail), *Sebastian Cabot, America*

MAXIMVS HIC DIES
MARE CONGELAT
CIRCVLVS
Terra Incognita
OCEANVS OCCIDENT
TROPICVS CA
MAR DEL SVR
Peru prouincia
TROPICVS
Terra uel mare incognitum
CIRCVLVS
MARE CONGELAT
MAXIMVS HIC DIES

PLATE 34
Sebastian Cabot (ca. 1476–1557)
World map
Antwerp, 1544
Copperplate engraving on 8 sheets, colored by hand, 50 x 84 in. (1240 x 2100 mm.)
Location:
Bibliothèque Nationale, Paris

THE "VALLARD" CHART, DIEPPE, BEFORE 1547

This highly decorative and practical mariner's chart is centered on the St. Lawrence area discovered by Jacques Cartier during his three voyages of 1534-42. After ascending the river on his first two expeditions of 1534-35, Cartier joined the Sieur de Roberval, a French nobleman, in an attempt to establish a colony in Canada. The "Vallard" chart is one of the first to map these explorations. It is embellished by exceptionally fine vignettes representing the earliest contacts of would-be French colonists with native Americans.

South is at the top, following the custom of Dieppe mapmakers who showed the northern and southern lands each viewed toward the equator. Emphasizing this arrangement are six bold compass roses from which rhumb lines radiate across the Atlantic.

Beginning at far north, the chart portrays "Labrador" (bottom left), actually Greenland, and the entrance to Davis and Hudson's Straits, as in earlier delineations. *Terra Nova*, our present Labrador and Newfoundland, are well drawn with many place-names, some in Portuguese from the voyage of João Alvarez Fagundes in 1520. From Cape Breton southward the coast is mapped to the Florida Cape. Here the chart uses earlier sources—the Ribero map of 1529 (Plate 29) and the Estebān Gomez reports. The Azores (*Illes des esores*) and Bermuda provide interesting points of reference at mid-ocean.

Of prime importance is *Rio do Canada*, representing new information from Cartier's voyages. The St. Lawrence River's vast estuary is considerably improved over earlier maps, although Nova Scotia awaits further development. Names on the river begin at *Totamagy* below the mouth of the Ottawa. *Ochelaga*, later Montreal, is shown on the north bank; Lake St. Peter is called *le Bac* (the ferry); and places known today such as Ile d'Orléans, Saguenay River, and Sept-Iles appear among the many villages marked along the northeastward course. The large island, Anticosti on modern maps, divides the river into two channels: today's Jacques-Cartier and the Gaspé Passages, which empty into the Gulf of St. Lawrence.

The map depicts colonists of Roberval's settlement established at *France-Roy* on the St. Lawrence River near Montreal. The survivors returned to France after one Canadian winter. Cartier's claim of the St. Lawrence valley for François I made him the father of French Canada, although more than a half-century elapsed before Champlain established the first permanent settlement.

The Vallard atlas, a beautiful and important cartographic treasure of the sixteenth century, was produced in Dieppe, the Norman seaport and center of fine navigational chart and instrument makers. The work was undoubtedly designed for a noble or royal patron, although the cartographer remains unknown; and no information exists on Vallard, the early owner whose signature gives the atlas its common name. After 1800, the atlas belonged to Prince Talleyrand and then passed to Sir Thomas Phillipps, famous nineteenth-century English book collector. This first map to focus on Canada and the American east coast is as artistic as it is historic.

References: Biggar 1911, pl. 10; Cortesão & da Mota 1960; Ganong 1964, 234 passim; Harrisse 1900, 228; Kohl 1869, pl. 19; Layng (comp.) 1956, no. 306; Lowery 1912, no. 45; Nebenzahl 1982; Winsor 1889, 4:87.

PLATE 35 (detail)
The "Vallard" Chart
Eastern Canada, oriented
with south at the top,
with first French settlers

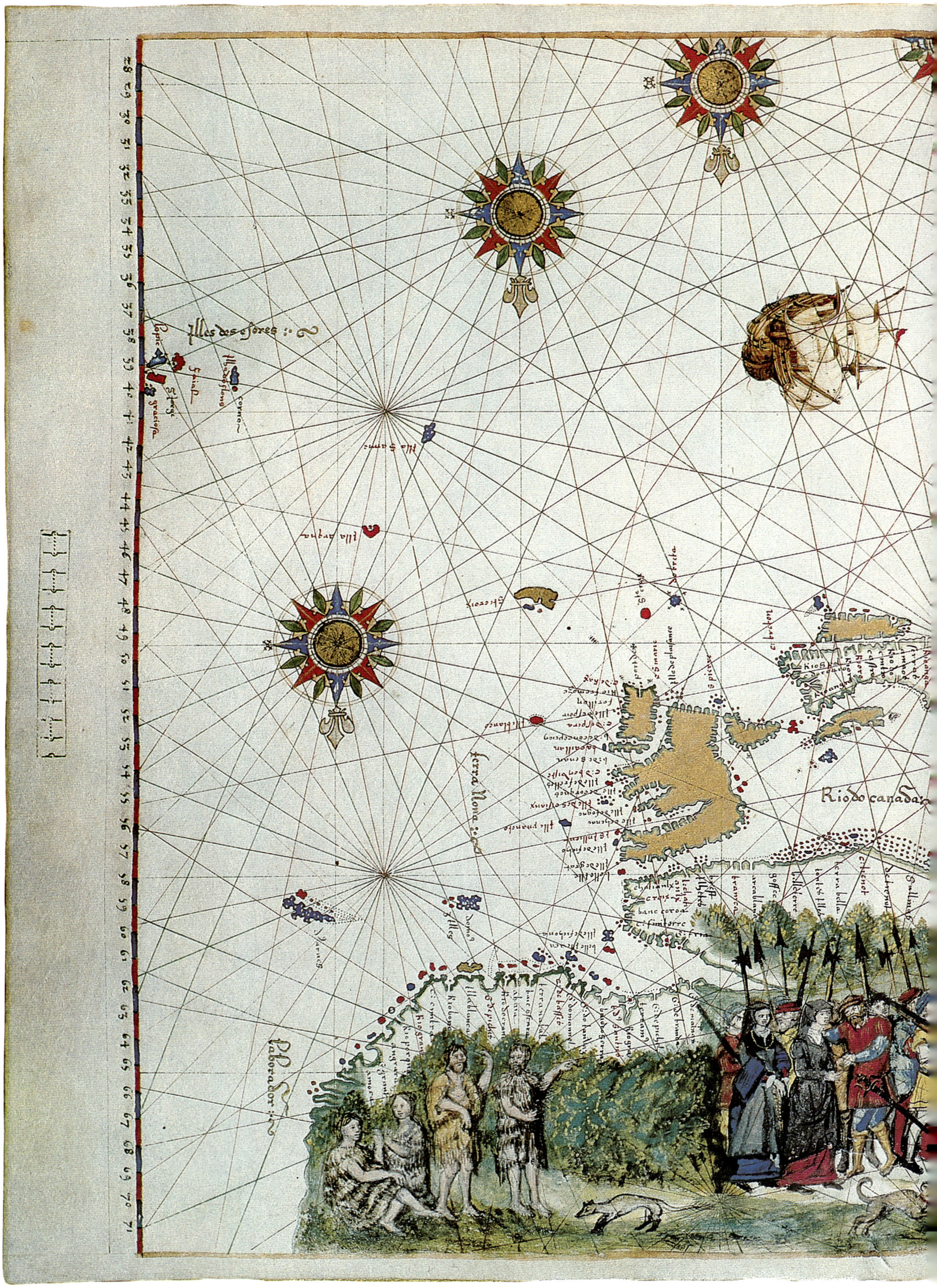
Illes des esores
Riodo canada

PLATE 35
The "Vallard" Chart
*East Coast of North America,
oriented with south at the top
Dieppe (France), before 1547
Illuminated manuscript
on vellum, 15.7 × 22.8 in.
(399 × 579 mm.)
Location:
Huntington Library,
San Marino, CA*

PIERRE DESCELIERS, WORLD MAP, ARQUES (DIEPPE), 1550

Pierre Desceliers, father of the French School of hydrography, produced this superb example of Renaissance chartmaking. Desceliers was born in the village of Arques outside Dieppe, a center of French maritime activity. A priest from a noble family of military men, he was a mapmaker, author of geographical works, and the first hydrographer commissioned to teach nautical science and to examine pilots at Dieppe.

Although the French were not the first great nation chronologically in mapmaking, by mid-sixteenth century, beginning with Desceliers and followed by others of the Dieppe School, they were producing the finest charts made in Europe. This fact coincided with France's belated entry into the contest for maritime supremacy. The Dieppe cartographers' charts grew out of the old coastal chart tradition, to which they added information about continental areas. Two orientations were employed: South of the equator the map appears "right side up," while in the north it seems "upside down." This is because place-names were designed to be read from the top or bottom border towards the equator.

Desceliers's elaborate illustrations, delicate and sensitive drawings, and rich coloration create a superb artistic rendition. Three coats-of-arms indicate that the chart was dedicated to Henry II, King of France, as well as to Grand Constable de Montmorency and Admiral Claude d'Annebaut. The compass roses and accurately portrayed ships are miniature masterpieces of design. Land areas are covered with flora and fauna derived from classical legends, biblical stories, medieval travelers, and geographers.

The minute detail of known coastlines contrasting with the wavy lines that complete the unexplored parts of continents combine to make a highly scientific delineation. America is carefully defined from Labrador to the Strait of Magellan, but the Arctic and Pacific coasts reveal the absence of empirical knowledge and show few specifics. An important new feature is the St. Lawrence River, recently discovered by Jacques Cartier and first brought to the world's attention by maps of the Dieppe School.

A decade after Verrazzano (Plate 28), Cartier was sent by the King of France in 1534 on another attempt to pierce the barrier of North America and to gain direct access to the rich markets of the Orient. Cartier's three voyages greatly enlarged Europe's understanding of what is now eastern Canada and laid the groundwork for future control by France. Explorations of this period led to discoveries and claims that had a profound effect on the subsequent languages and cultures of the various countries to emerge in the New World. Spanish-speaking American nations from Mexico to Chile, Portuguese Brazil, French Quebec, and later Anglo-American United States all trace their roots to the colonies established by the European countries who first introduced their customs to these lands.

Desceliers's famous landmark in cartography, with its great size, beauty, and current information on new discoveries, allows us to see how royalty and nobility were informed of the expansion of Europe's interest in the rest of the world.

References: Biggar 1911, pl. 14; Ganong 1964, 233 passim; George 1969, 24 passim; Hoffman 1961, ch. 13; Layng (comp.) 1956, no. 323; Mollat du Jourdin 1984, no. 47; Schwartz & Ehrenberg 1980, 55; Skelton 1958, 93.

PLATE 36 (detail), *Pierre Desceliers, The New World*

REGION FROIDE
CANADA
MER DE FRAN CE
MER DESPAIGNE
MER OCCEANE
MER DES ENTILLES
EVROPE
LE PERV
LA LIGNE
LA ZONA TORRIDA
MER DV SV
AMERIQVE
TROPIQVE DE
MER AVSTRALLE
MER DE MAGELLAN
LA TERRE

PLATE 36 *Pierre Desceliers (1487–1553), World map, Arques (Dieppe, France), 1550
Illuminated manuscript on 4 vellum leaves made to be joined,
53 x 84 in. (1350 x 2150 mm.) Location: The British Library, London*

GUILLAUME LE TESTU, EAST COAST OF NORTH AMERICA, FLORIDA, & THE GREATER ANTILLES, LE HAVRE, 1556

The Dieppe school of cartographers that flourished in Normandy at mid-sixteenth century created a number of beautiful atlases, but none more striking than the *Cosmographie Universelle* of Guillaume Le Testu, the colorful Norman-Huguenot hydrographer. His coastlines have the precise detail required by navigators, while inland the natives and their dwellings, the trees, and animals are rendered with considerable charm. The oceans, decorated with drawings of contemporary ships, have wave patterns that provide the unique seascape quality characteristic of Le Testu's work. His 1556 map of Florida includes an area he himself was soon to explore with the French Huguenots. The map is derived from Spanish and Portuguese sources but includes place-names from Verrazzano's 1524 French voyage.

Born around 1512 at Le Havre, Le Testu first became a pilot at Dieppe, then chief pilot of his native port at the mouth of the Seine. During the period between Cartier's voyages to Canada in 1534-42 and Champlain's first voyage in 1603, France's interests in America focused on Brazil and Florida. Le Testu sailed in 1555 with the first Huguenot colonizing expedition sent by Admiral de Coligny. That same year he dedicated the atlas containing this map to the admiral.

After the French Brazilian settlement failed under his most trusted captains, Ribaut and Laudonière, Coligny dispatched more Huguenots to the east coast of Florida. The Spanish had not as yet settled Florida, and it would be another two decades before the English became interested in Roanoke. The French attempted to plant colonies at Port Royal (South Carolina) and at Fort Caroline near the mouth of St. John's River, Florida.

More than national sovereignty determined the success or failure of these early colonies, however. Colonists from Spain and France periodically were massacred with a religious fervor reserved by Spanish Catholics and French Protestants for each other. Founding the port city of St. Augustine, in 1565, finally enabled Spain to establish its authority on the mainland, protect shipping in the Bahamas Channel, and dispose of the French incursion.

In 1572, Le Testu was given command of a ship to reconnoiter Nombre de Dios, the Caribbean port of Panama from which Peruvian gold and silver was transshipped to Spain. There he joined the Englishman, Francis Drake, in an attempt to hijack the mule train carrying the immensely rich cargo. They were attacked by Spanish soldiers. Drake, though wounded, escaped to live out his historic career, but Le Testu was killed by a shot from an arquebus (early rifle). Thevet, famous French writer on voyages, mourned the death of "one of the most expert pilots of the age."

The colorful figures of Admiral de Coligny and Guillaume Le Testu, the magnificent atlas Le Testu dedicated to the admiral, and this map of Florida that it contained are tied together in a tragic chapter of colonial history from which France did not recover. There was no further French activity in North America, south of Canada, until LaSalle's attempt to found Louisiana over a century later.

References: Cumming, Skelton & Quinn 1972, no. 163; Hoffman 1961, ch. 13; Mollat du Jourdin 1984, no. 48.

PLATE 37
Guillaume Le Testu (ca. 1509-72)
East Coast of North America, Florida, and the Greater Antilles
Le Havre (France), 1556
Illuminated manuscript on paper,
15 x 10.5 in. (370 x 265 mm.)
Location:
Ministère de la Défense
Service Historique de l'Armée de Terre,
Vincennes (France)

TERRE DE LA FLORIDE

PARTIE DE LA MER OCEANNE

LA COVBE

ESPAIGNOL

PARTIE DE LA MER DE LENTILLE

PLATE 38
Diogo Homem (fl. 1530–76)
The North Atlantic
London, 1558
Illuminated manuscript on vellum, 24 x 33 in. (590 x 820 mm.)
Location: The British Library, London

SEPTENTRIO

Desertú bus o2
Terra agricule.

Islanda

Mare ibernicú.

Mare terre noue.

Mare occanum:~

Mare galicú

Mare hispanie.

Insule aroreas.

S OCCIDUUS.

Insules fortunate.

ORIENS

TROPICVS CANCRII

J. p. uiridis.

Canagua R.

DIOGO HOMEM, THE NORTH ATLANTIC, LONDON, 1558

Portugal, the small nation sharing the Iberian peninsula with Spain, began Europe's age of overseas discovery with a series of voyages starting around 1420. Continually probing the Atlantic coast of Africa, Portuguese navigators by the end of the century had rounded the Cape of Good Hope and opened the gateway to the East. At mid-sixteenth century, Portugal remained one of Europe's chief maritime powers, dominating trade with Brazil, India, and the Spice Islands. At the same time the tiny nation avoided competition with Spain over North America and the West Indies.

Although the French port of Dieppe had emerged as the center of fine chartmaking during the 1540s (Plates 35 through 37), Portuguese cartographers continued to excel in this increasingly sophisticated art and science. The Homem family contributed greatly to Portugal's continued leadership in hydrography, beginning in 1517, when Lopo Homem was made "Royal Master of Sea Charts." Initially assisted by his former teacher Pedro Reinel, Homem and his sons Diogo and Andreas left a remarkable body of individual sailing charts and maritime atlases produced throughout the century.

Diogo Homem became the most prolific Portuguese cartographer of the 1500s. Although many facts of his life remain unknown, he and his brother maintained the family tradition of serving royalty. Andreas became mapmaker to the King of France, while Diogo was commissioned in London to produce the beautiful atlas in which this chart appeared for Queen Mary of England and her husband Philip II of Spain. It was historically altered in 1558 when the Catholic Queen died and Philip's coat-of-arms on the map of England was obliterated, no doubt to satisfy Elizabeth, Mary's piously Protestant successor. Homem was living as an exile in England at the time, having been suspected of committing a murder in Lisbon. Although later exonerated, he never returned to Portugal.

The Atlantic chart illustrated here, enclosed within a checkered border indicating degrees of latitude and longitude, includes at right, Iceland, Ireland, western Spain, Portugal, and West Africa from the Strait of Gibraltar to Cape Verde. The West Indies are portrayed to Grenada with the Greater Antilles particularly well delineated. The Guajira Peninsula (Columbia), Cabo Gracias a Dios (Honduras), and Cabo Catoche (Yucatán) enclose the Caribbean Sea.

Improved mapping of the Bahamas, the Florida cape, Cuba, and Yucatán clarifies the channels necessary for West Indian navigation. Homem's Carolina coast follows the earlier Spanish *padrón general*, while the area from the mouth of the Chesapeake to Maine is telescoped, reflecting charts of returning explorers who circumvented much of this coast. Nova Scotia, however, is greatly enlarged, the result of an enormously magnified Bay of Fundy. Although the bay was discovered by Fagundes while he was attempting to found a Portuguese settlement in the 1520s, it was just beginning to appear on maps.

As on other contemporary charts, the interpretation and delineation of the St. Lawrence Valley and Gulf presented great difficulty. The river's north shore is shown as an archipelago with intervening inlets, any of which might provide the legendary northwest passage to the Pacific. The western ocean (Pacific) is named by Homem *Mare le Paramatiu* after the Dieppe navigators, the Parmentier brothers, who were the first Frenchmen in those waters.

This elegant chart, commissioned by the English crown, bridges the gap between Spain and Portugal's official maps of the first half of the century and the great cartographic improvements that began in the Low Countries during the 1560s.

References: Cortesão & da Mota 1960, pl. 106; Cumming 1988, 22, no. 14; Cumming, Skelton & Quinn 1972, no. 140; Ganong 1964, 74ff; Hoffman 1961, ch. 15; Kohl 1869, no. 21; Layng (comp.) 1956, no. 423; Nordenskiöld [1897], 67; Winsor 1889, 2:227 passim.

ABRAHAM ORTELIUS, WORLD MAP, ANTWERP, 1564

Abraham Ortels, known as Ortelius, initially a map seller and colorist of Antwerp, was second only to his friend and rival, Gerardus Mercator, as a leader of the cartographic renaissance in the Low Countries during the late sixteenth century. Ortelius chose for his first major mapmaking effort this heart-shaped projection to display the spherical world on a flat surface. His configuration eliminates the pointed bottom of the heart and is known as a "truncated" cordiform. Like most great world maps of the period, it combined accurate new information—and occasional misinformation—with erroneous ancient beliefs. In 1564, Ortelius boldly issued this new map on eight sheets, which when joined made a wall map five feet long.

He rejected the continental masses in the Arctic and the "second Greenland," still included on Mercator's map (Plate 40). However, his unusual projection dramatically emphasized the great Antarctic continent, a holdover from classical Greek cosmography. The East Indies are accurately mapped for the period, although earlier Portuguese manuscript charts had improved the contours of Sumatra and Java. For Europe, Africa, and Asia, Ortelius drew upon the fine maps of the Venetian cartographer, Giacomo Gastaldi.

Maps such as Gastaldi's multiple-sheet renditions of Asia and Africa, and especially his recently discovered 1561 wall map of the world, had set the standard of excellence for the Italian school of cartography. This school led the field just before Ortelius, Mercator, Hondius, and Blaeu combined their talents to capture that leadership for the Low Countries.

Ortelius's North America on the heart-shaped world shows a fundamental misconception of the complex Gulf of St. Lawrence region and Canadian maritime provinces. His Labrador, Terra Nova, Canada, Norumbega, and Nuova Franza form an archipelago beginning too far eastward in the North Atlantic. A half-century earlier when the "land of the Cortereals" first appeared on maps representing the Portuguese discovery in Labrador, it was located in the northeastern Atlantic disconnected from the North American continent, then called Florida. Ortelius was attempting to tie these elements together, but his solution was seriously flawed.

Northern North America in the area of New France terminates with the south shore of the St. Lawrence River, leaving the territory to the north (most of Canada) represented as part of the Arctic Ocean. To the west, the famous fugitive land of Anian is placed in northeastern Asia, across the strait that separates that continent from North America; the name Sierra Nevada appears in northern California's Quivira region. The Strait of Anian first appeared on Gastaldi's long-lost 1561 nine-sheet woodcut world map that surfaced in 1978. Ortelius's continental mountain ranges of North America run incorrectly east-westerly along the 40th parallel.

At lower right, carefully drawn insets present Cuzco and Mexico City based on views recently published in Venice by Ramusio, indicating the degree of interest in Spain's colonial activities. If anyone questioned the motivation behind ambitious cartographic projects such as this, Ortelius provides a suitable rationale in the large table at lower left, where his text lists worldwide sources of gold, silver, precious stones, and spices. Within the clouds surrounding the map, twelve finely drawn Flemish faces represent the winds, identified by their Latin, Italian, and Dutch names. At bottom left Gerard De Jode is recorded as the publisher.

Ortelius's large map signified the migration of cartographic publishing from Italy to the Low Countries. Although this work was eclipsed by Mercator's world map of 1569 (Plate 40), a year later Ortelius published the first modern atlas. Called the *Theater of the World*, it was the earliest atlas in which the contents were determined by the publisher rather than by the purchaser. Many editions followed, as did rival works by De Jode and Mercator. For the next 150 years, first Antwerp, then Amsterdam, became the world center of geographical publishing.

References: Bagrow/Karrow 1990, no. 1:1; Ganong 1964, 41213; Hoffman 1961, ch. 15; Layng (comp.) 1956, no. 473; Schilder 1986-88, 2:pl. 1; Shirley 1983, no. 114; Wagner 1937, no. 63.

NOVA TOTIVS TERRAR
TERICORVM TRADIT
ABRAH. ORTELIO
ANVERPIANO AVCT.
CORVS
Ponente maestro
CIRCIVS
Maestrale
OCCIDENS
Ponente
AFRICVS
Ponente libeccio
LIBO NOTVS
Libeccio
MERIDIES
Mezo
ZONA TEMPERATA
ZONA TORRIDA
ZONA TEMPERATA
ZONA
OCEANVS OCCIDENTALIS
uulgo
MAR DEL NORT
MARE MAGELLANICVM
SIVE PACIFICVM
uulgo
MAR DEL ZVR
AEQVATOR SIVE AEQVINOCTIALIS
TROPICVS CAPRICORNI SIVE BRVMALIS
CIRCVLVS ARCTICVS
CIRCVLVS ANTARCTICVS
PERV
TERRA DEL FVEGO
MERCIVM ALIQVOT PEREGRINARVM, ET LOCORVM VNDE AD NOS ADFERVNTVR, CATALOGVS
NOBILI ET ERVDITO MAR
LAVRINO D. DE WATERVL
ABRAHAMVS ORTELIVS
ANVERPIANVS
D. D.
Prostant Antuerpiae, apud Gerardum de Jode; in Borsa noua
Cum Regiae Maiestatis Priuilegio ad sexennium

PLATE 39 *Abraham Ortelius (1527–98)*, Nova Totius Terrarum Orbis...., Antwerp, 1564
Copperplate engraving, 8 sheets, each 15 x 17 in. (375 x 440 mm.)
Location:
The British Library/Map Library, London

PART IV

Europe's Colonial Era Begins

When Columbus founded the first European settlement on Hispaniola in 1493, he ushered in the European colonial epoch in America. By the time Hernando Cortes launched his conquest of Mexico in 1519, the Spanish had established colonies on Cuba, Puerto Rico, and Jamaica. Cortes captured the Mexican capital two years later from Montezuma and the Aztecs. The details of this amazing city are carefully mapped on a remarkable woodcut that accompanied Cortes's letter to Charles V, published in 1524. It appears along with a Gulf of Mexico chart illustrating Cortes's narrative of the tragic voyage of Alonso de Pineda.

In 1533, Francisco Pizarro and a handful of men seized Cuzco, the capital of Peru, in the face of an Inca army reputed to number 30,000. Included in this section of the atlas is a contemporary detailed plan of Cuzco, which also depicts the Inca emperor being carried on a royal litter. This part contains superb delineations of the Spanish strongholds at Cartagena, Santo Domingo, and St. Augustine, published in celebration of Sir Francis Drake's successful campaign against these cities.

Part Four begins with one of the most notable maps of the Renaissance—Gerardus Mercator's 1569 world chart. It is drawn on the projection bearing his name, which is still in use today. Mercator's innovations were adapted by such important mapmakers as Abraham Ortelius and Cornelis De Jode, whose maps in turn were widely emulated. The Mercator world map also influenced the geographical thinking of the early advocates of a British Empire. England was dependent on the best charts from the Low Countries, where mapmaking was more advanced than elsewhere in Europe.

One of the most outstanding plates in this section is the Hondius-Drake double-hemispheric map displaying the track of Sir Francis Drake's voyage around the world. Drake proved on this expedition that his seamanship was equal to that of Columbus and Magellan. This chart displays important refinements in mapping South America for which Drake was responsible.

The Elizabethans had several designs on North America. Drake had reached California, Martin Frobisher was searching for a northwest passage in the Canadian arctic, and Sir Walter Raleigh assumed the patronage of the Virginia enterprise. The British founded their first colony on Roanoke island, north of Spanish-controlled territory. One reason for selecting this site was the long-held hope that Verrazzano had been correct when he reported that the waterway just west of the Carolina Outer Banks connected with the Pacific Ocean. European dreams of a passage to the Orient remained alive for some time. The two superb watercolor maps of Virginia by John White and Thomas Harriot are one of the fruits of Raleigh's ill-fated colonial attempt.

Part Four, which begins with the Mercator world map, ends with the first practical application of Mercator's projection on Edward Wright's world chart (Plate 50). An Elizabethan mathematician, Wright described how to use the projection, which led to its widespread adoption. The historic maps in this section reflect the permanent establishment of European influence in America.

Gerardus Mercator, the leading Renaissance mapmaker, and his successor, Jodocus Hondius, depicted with the accoutrements of their trade.

The famous 1583 portrait of Sir Francis Drake, issued in the eighteenth century by George Vertue.

GERARDUS MERCATOR, WORLD MAP ON MERCATOR'S PROJECTION, DUISBURG, 1569

The foremost Renaissance mapmaker was Gerhard Kremer, a native of Rupelmonde, Flanders, who was known as Gerardus Mercator. Despite his family's modest circumstances, Mercator attended the University of Louvain, where he received a master's degree at age twenty. He became a protégé of Regnier Gemma Frisius, the leading mathematician and astronomer in the Low Countries and founder of the Belgian school of geography. After learning Frisius's sciences, Mercator studied engraving and calligraphy, developing and refining these skills that would serve him so well in creating his great works.

Mercator's comprehensive preparation, innate genius, fine italic handwriting, engraving ability, and the intellectual atmosphere of Louvain all contributed to his early success as a map, globe, and scientific instrument maker. By twenty-five he had produced a remarkable six-sheet map of Palestine, and the following year, a double heart-shaped map of the world. His productive career lasted past his eightieth birthday.

His most important contribution was this 1569 world chart, drawn by a new method of his own invention. Universally known as the Mercator projection, its parallels and meridians appear as straight lines intersecting at right angles to produce at any point an accurate ratio of latitude to longitude. He accomplished this feat by employing a formula that gradually increased the length of latitudinal degrees from the equator to the North Pole. For the first time navigators had a chart with true directions on which a ship's course could be plotted on a straight line.

Although his world map is considered a masterpiece, for many years navigators' skepticism of any innovation—coupled with Mercator's failure to publish the formula he used to develop his work—prevented practical adoption of the projection. Thirty years later, after Mercator's death, Edward Wright wrote a treatise, *Certaine Errors in Navigation*, explaining the projection and converting it into a feasible, working navigational aid. After Wright's interpretation and examples, the Mercator Projection slowly became the chart that navigators preferred. Today, over three centuries later, though somewhat modified, Mercator's projection still serves the purpose for which it was devised.

Considering the state of knowledge at the time, the map is a paradox of advances and retrogressions. An enormous Antarctic continent is separated from South America by only the Strait of Magellan and possesses undefined large promontories extending into the East Indies. In the Arctic is another "continent" divided by four huge rivers running toward the pole from the Arctic Ocean. The North Polar inset at lower left attempts to justify this conception. It is based upon the fourteenth-century legend of a circular landmass divided into four sections by great rivers discharging into a sea at the North Pole. Mercator's Arctic Ocean encouraged thoughts of northwest and northeast passages from Europe to China and the East Indies that stimulated further exploratory voyages.

In North America there is a relatively accurate peninsular-shaped lower California, but a huge protuberance distends southwest South America (Drake corrected this misconception on his circumnavigation ten years later). The discovered parts of the St. Lawrence River are well represented from Cartier's reports, and Greenland is separated from Canada. Old legends die hard, however, and a false second Greenland appears in the north, as do nonexistent islands in the Atlantic reported by the Zeno brothers in the fourteenth century.

From Florida to Labrador (*Terra Corterealis*), Mercator has followed Spanish, French, and Portuguese sources, but he omits Verrazzano's information. Mercator's New England and Nova Scotia coast extends too far east-westerly, a problem in all early maps caused by erroneous magnetic compass readings at that latitude. However, the Appalachian Mountains first appear as a continuous range running parallel to the Atlantic coast. The entire delineation became the prototype followed by mapmakers into the beginning of the next century. Ortelius of Antwerp, whose maps and atlases sold widely throughout Europe, published the first edition of his atlas in 1570. His world and America maps were derived from this Mercator chart of 1569.

The extraordinary wall map on which Mercator unveiled his remarkable projection has survived in only a few copies. Along with its many ornamental designs and allegorical figures, it represents one of the most significant Renaissance scientific accomplishments. Mercator spent decades studying every projection known to cartography. He was aware of what mariners required; and by extensive and tireless application of his skills and training, he invented a solution that remains in use today.

References: Bagrow/Karrow 1990, no. 56:17; Boorstin 1983, 273-78; Brown (comp.) 1952, no. 132; Cumming, Skelton & Quinn 1972, 98; Fite & Freeman 1926, no. 22; Ganong 1964, 415-27; Shirley 1983, no. 119.

PLATE 40 (detail)
Gerardus Mercator
America
Location:
Maritiem Museum Prins Hendrik, Rotterdam

NOVA ET AVCTA ORBIS TERRA

PLATE 40 *Gerardus Mercator (1512–94)*, Nova et Aucta Orbis Terrae Descriptio, *Duisburg (Germany), 1569*
Copperplate engraving on 18 sheets, 52 x 78 in. (1325 x 1950 mm.)
Location:
Bibliothèque Nationale, Paris

GEORG BRAUN & FRANS HOGENBERG, PLAN OF CUZCO, COLOGNE, 1572

Cuzco (Cusco), over two miles high in the Andes, was the capital of the Empire of the Sun, a larger and richer civilization than the Mayan. Ruled by the Incas, it was one of the most advanced native American cities. At its peak, before the conquest in 1533 by Pizarro, the Inca empire—centered at Cuzco—ranged from Ecuador to Chile and from the Andes to the Pacific, with a population estimated at eight to twelve million.

After Balboa discovered the Pacific in 1513, Panama became headquarters for intrigue, where Europeans hatched plots of South American adventure. Panama served as the base for the conquest of Peru as Cuba had for Mexico. Pizarro and his fellow *conquistadores*, after several false starts, launched their invasion in 1532. Three years later the Incas were subdued, and Pizarro founded Lima, "City of the Kings," beginning the end of Cuzco's dominance.

In his distinguished collection of narratives relating to America, notable for its accuracy, Giovanni Battista Ramusio in 1556 accompanied the account of Pizarro's conquest of Peru with this plan of the Inca capital. Based partly on fact and partly on the artist's imagination, it contains many elements of the city. The image was popular in Europe for over fifty years, appearing in many editions of several different works. This illustration and one of Mexico are the only American subjects in *Civitates Orbis Terrarum* (Cities of the Lands of the World) by Georg Braun and Frans Hogenberg. The figures in the foreground of Indians bearing the Inca ruler was added for this engraving.

Two famous streams, the Huatenay and Rodadero, can be seen on either side of the right-angle grid town with its great plaza before the Temple of the Sun of the Incas (today, partly occupied by the convent of San Domingo). The famous cathedral of Cuzco was constructed later. Hardly a city of grass shacks or tepees, Cuzco amazed the Spaniards by its solid masonry construction. So massive are the cut-stone foundations that historians and engineers are still mystified by how the Incas put them in place. The plain at right leads to the legendary hill of Sacsahuaman with its ruins of the main Inca fortress. European readers fascinated by accounts of the fabulous capital, captured by the Spanish, had only this view to help them imagine how the greatest city in the New World appeared.

References: Brown (comp.) 1952, no. 269; Skelton 1965a, xliii.

PLATE 41

Georg Braun (1541–62)
& Frans Hogenberg (1535–90)
Cuzco Regni Peru
in Novo Orbe Caput
Cologne, 1572

Copperplate engraving, colored by hand. 10.5 x 9.3 in. (268 x 235 mm.)

Location:
The Newberry Library, Chicago

CUSCO

CVSCO, REGNI PERV IN NOVO ORBE CAPVT.

JODOCUS HONDIUS, WORLD MAP, LONDON, CA. 1589

The world surmounted by Queen Elizabeth's royal coat-of-arms is a reflection of the imperial spirit rising in England during the 1580s. The circumnavigations of Sir Francis Drake in 1577-80 and Thomas Cavendish in 1586-88 boosted the morale of Elizabethan England. These first English round-the-world voyages, traced on this map, proved highly successful. They not only returned many times their investors' advances, but they represented the kind of daring exploit and accomplished seamanship that propelled England into the forefront of European overseas powers.

Drake's expedition was achieved with the boldness and relish for action that made his name a household word in seaports of the world. He was intensely admired by friends and mortally feared by his enemies. It remains unknown whether the main purpose of his world voyage was to discover the long-sought northwest passage, sack Spanish ports and attack shipping, establish trade with the Spice Islands, or found a colony. Despite losing four of his five original vessels, Drake brought home huge quantities of gold and silver captured in South America and many casks of Molucca spices. The only cargo lost was jettisoned to help lift the ship off the rocks during a crisis at Celebes in the East Indies (depicted at bottom right).

His outward voyage seemed interminable, taking nine months from home port to the Pacific. When he emerged from the Strait of Magellan, two months of violent storms blew him southward where he inadvertently discovered Cape Horn. He then sailed north up the coast of South America. After pillaging the Spanish ports, including Callao (Lima's harbor and key to the riches of Peru and the Incas), Drake captured a large Spanish vessel laden with bullion. The *Golden Hind,* only surviving ship of Drake's fleet, now loaded with all the gold and silver she could carry, continued northward. Drake attempted unsuccessfully to locate the Pacific side of a northwest passage and possibly to rendezvous with Frobisher coming from the Atlantic.

Drake became the first European to visit the California coast. He spent many weeks there, took possession for England, and named it New Albion. He claimed that before he embarked for the East Indies, the California Indians made him their king. This has been interpreted as a first attempt at English colonization in North America. The English, however, lacking a northern sea route, never returned.

Upon arriving in the Moluccas, Drake learned that the natives and the Portuguese had quarreled, giving him a propitious moment to trade for spices. At Ternate, the Sultan had four war canoes tow the *Golden Hind* to port (depicted at bottom left).

Cavendish's voyage was less dramatic, taking two years and losing two of three ships but also bringing home a profitable cargo of looted Spanish bullion and Molucca spices. Geographically, except for locating a new harbor in southern Patagonia, Cavendish contributed little. Later he died at sea at the age of thirty-one attempting another circumnavigation.

Jodocus Hondius, a Flemish engraver living in London, probably produced this map before moving to Amsterdam in 1593. He later headed the great firm of cartographical publishers founded by Mercator in the Netherlands that carried on for four generations. To draft this double-hemispheric map, Hondius used in simplified form a combination of available Mercator and Ortelius delineations. He divided the two hemispheres to display the prime meridian at mid-Atlantic and mid-Pacific for clarity in tracking the routes.

When a very pleased Queen Elizabeth honored Drake by having the *Golden Hind* permanently moored in the Thames, one of its visitors was Hondius, who made the likeness that appears at bottom center. The top vignettes show Drake's exit from Java and, most controversial of all, his California port. The site of his landing has been long argued by Drake experts, with passionate proponents variously proposing San Francisco Bay, Drake's Bay, and Trinidad Bay.

Drake's legends and deeds promoted England's interests. This map became an icon representing Britain's first circumnavigation and indicating Elizabeth's intention to make England a world power.

References: British Museum, 1977, no. 78; Fite & Freeman 1926, no. 27; Power; Shirley 1983, no. 188; Skelton 1958, nos. 90 & 91; Sprent 1927, 2; Thrower (ed.) 1979, no. 79; Wagner 1926, 417-19; 1937, no. 176.

PLATE 42 (detail). *Jodocus Hondius, the Pacific Ocean and California*

VERA TOTIVS EXPE

Descriptio D. Franc. Draci qui 5. navibus probe instructis, ex Anglia solvens 13. Decembris
ceteris partim flammis, partim fluctibus correptis, in Angliam rediit 27 Septembris 158
Angli, qui eundem Draci cursum fere tenuit etiam ex Anglia per universum orbem; sed
quinto Septembris 1588. in patriæ portum Plimmouth, unde prius exierat, mag

Portus Novæ Albionis

Gilolo In.

Non immerito, amice lector, formam navis F. Draci
huic nostræ tabulæ adjungi putavimus, miraculi enim
videri poterit; non solum hujus magnitudinis navi
& 20 hominum spacio, in scopulum illifi, onusta præter
ea, auro & argento, etc. posse tantum iter perfici, scil.
ad minimum, 8500 milliarium Germanicorum
Servatur in Anglia etiamnum navis illa, perpetuæ
memoriæ causa, Debfordiæ ad Tamisim, vale.

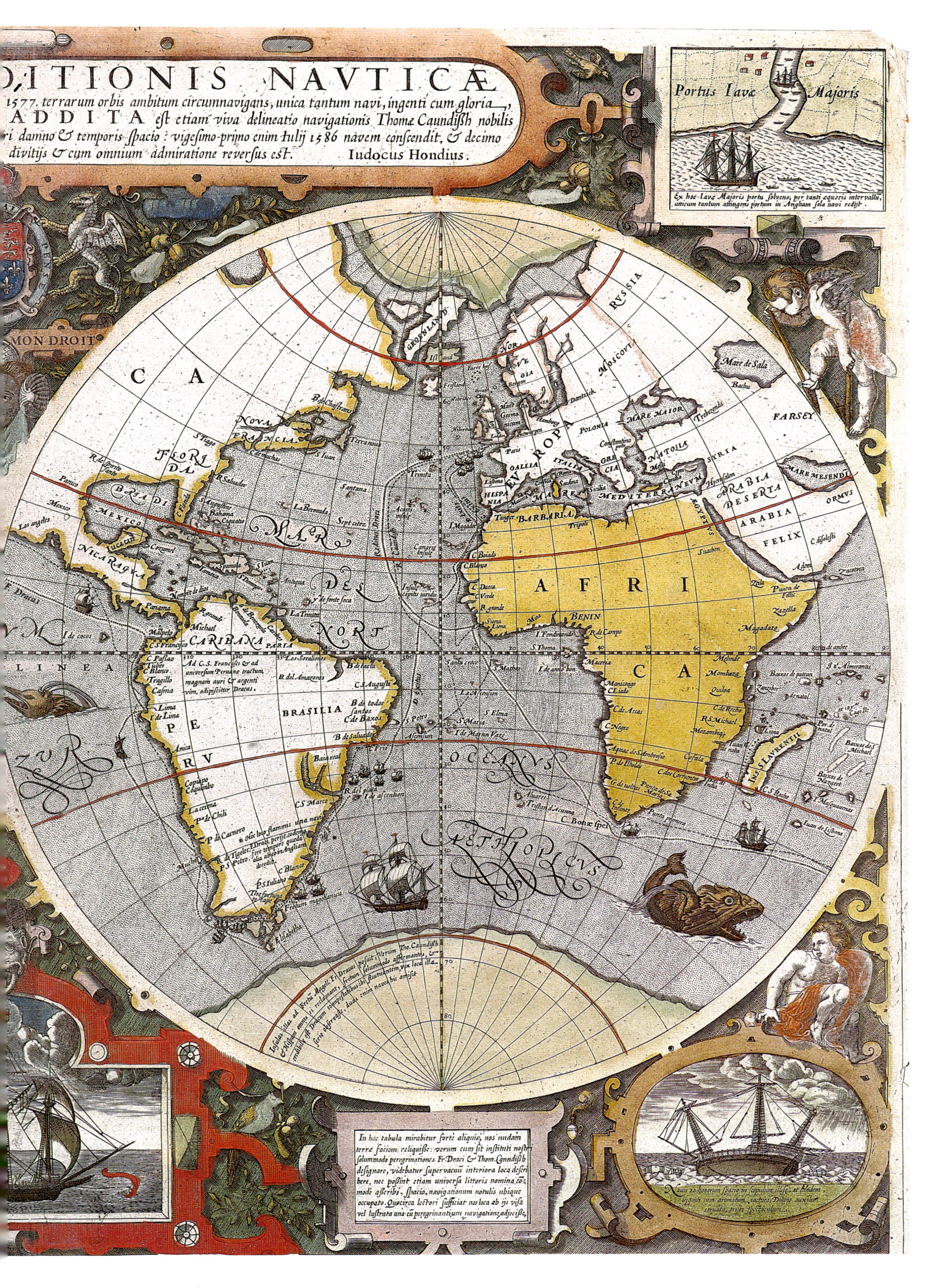

PLATE 42
*Jodocus Hondius
(1563–1612)*
Vera Totius Expeditionis
Nauticae Descriptio....
*London, ca. 1589
Copperplate engraving,
colored by hand,
12 x 18.5 in.
(305 x 470 mm.)
Location:
Robert H.
and Margaret C. Power
collection*

135

JOHN WHITE, *LA VIRGINEA PARS* & *LA VIRGENIA PARS*, LONDON, 1585–86

The English started late in Europe's race for American territory while Spain, Portugal, and then France were making their claims. Attempts to find a northwestern route to Asia began with the shadowy voyages of John Cabot in the 1490s and the enigmatic expedition of his son Sebastian around 1508. These early English probes produced no tangible results. In the 1570s, Martin Frobisher tried three times to discover a northwest passage. Although his explorations were useful, again no permanent benefits resulted. Other than raids on Spanish West Indian shipping by the slaver-corsair Hawkins brothers and their cousin Francis Drake, the English played no active role in America before Sir Walter Raleigh's first Virginia colony.

Drake's circumnavigation of 1577-80 fired the imagination of Englishmen and encouraged them to establish an English foothold on the North American continent. Elizabethan England's attention was focused on resources to be exploited, the need to establish a strong Protestant force to counterbalance Spain's Catholic American empire, the divine mission to deliver the faith to the "savages," and not least, to secure bases from which to raid Spanish sea lanes.

Sir Humphrey Gilbert conceived the idea of a Virginia colony along the vast unoccupied coast between Florida and Cape Breton. He obtained a patent from the Queen in 1578 to establish the first English colony. Unfortunately, his plans did not come to fruition, with the exception of his taking possession of Newfoundland in 1583, the year he died at sea. Raleigh, Gilbert's half-brother, then took over as promotor of the Virginia colony.

These two manuscript maps, drawn from surveys by the first English colonists, were the most accurate delineations of sixteenth-century North America. They resulted from the wise selection by Raleigh of two outstanding members of the expedition, Thomas Harriot and John White. Harriot, chronicler of the Roanoke colony, was an accomplished mathematician who understood navigating and surveying instruments and was undoubtedly responsible for the precision of the map of the Carolina banks and sounds. John White, the expedition's artist and later governor, drew the finished maps from field surveys. His charts and drawings of the natives and natural history subjects miraculously survived the centuries, fire, and flood. They provide a rare and important pictorial record of the first Elizabethan colonial effort.

White's general map of the southeast is a synthesis of the North American cartography that preceded him, augmented by his survey of the Roanoke Island region. By combining maps, he mixed his distance scales, as the oversized Bahamas demonstrate. His sources, identified by the historian David Quinn, include the Spanish chart brought by the expedition's Portuguese pilot, Fernandes; Jacques Le Moyne's map of Florida; a Dieppe chart from mid-century showing a waterway leading from Port Royal to the Sea of Verrazzano (Pacific); and an unknown Spanish map for Florida's west coast. White's chart is a summary of the Roanoke colonists' geographic knowledge of North America below the Chesapeake.

The second map by White is the first English survey drawn in America. It provides the northern portion of the general map and for the first time gives precise detail of the area from the entrance to Chesapeake Bay to south of Cape Lookout. Earlier Spanish charts were discarded, and the mathematical and surveying skills of Harriot, combined with White's draftsmanship, make this the most important surviving map of the early colonial period. Nevertheless, the further the map extends from the colony on Roanoke Island, the less accurate it becomes.

The off-setting visible on both maps results from the fire at Sotheby's auction gallery, London, in 1865 and the consequent water damage to which White's album of drawings was exposed.

Preparations during 1585–88 for the war with Spain and the Battle of the Armada contributed strongly to the failure of the Roanoke colonies. They survived long enough, however, to give the English sufficient confidence to sustain their imperial impulse. A few years later, John Smith founded another Virginia colony on the Chesapeake that was to give England its first real foothold in the New World.

References: Cumming 1962, nos. 7 & 8; 1988, ch. 8; Cumming, Skelton & Quinn 1972, nos. 199 & 213; Fite & Freeman 1926, no. 26; Hulton 1984, 32-34 passim; Hulton & Quinn (eds.) 1964; Morison 1971, ch. 4; Parry 1963, ch. 13; Penrose 1955, ch. 12; Quinn 1955, 1:460-61; 2:848; Skelton 1958, ch. 12; Waters 1955, 180; Wroth 1970, no. 53.

PLATE 43
John White (fl. 1577–93)
La Virginea Pars
London, 1585–86
*Illuminated manuscript on paper,
19 x 9 in. (483 x 229 mm.)*
Location:
The British Museum, London

138

PLATE 44
John White (fl. 1577–93)
La Virgenia Pars
London, 1585–86
Illuminated manuscript on paper, 14.5 x 18.5 in. (368 x 470 mm.)
Location:
The British Museum, London

BAPTISTA BOAZIO, *THE FAMOUSE WEST INDIAN VOYADGE...*, LEIDEN, 1588

Baptista Boazio, an Italian mapmaker living in London, is remembered for his series of five engravings celebrating Sir Francis Drake's raid on Spain's fortified West Indian cities. Of the series, four are city plans, including Santiago in the Cape Verde Islands, Santo Domingo on Hispaniola, Cartagena on the South American Coast, and St. Augustine in Florida. The fifth is this map with the track of the voyage.

It displays the Atlantic theater with the American coast from Labrador south to Rio de la Plata, where a quotation from Drake indicates that "seventien or 18 degrees to the Southwarde lye the Straites of Magellanos." The Caribbean and Gulf of Mexico are particularly well drawn and ironically were derived from a Spanish source, the official world chart maintained at Seville. Geo-political divisions are reflected by national flags: France's pennant in Canada, the English banner in Virginia, and Spanish colors over Latin America (Portugal had fallen to Philip II in 1580).

Jodocus Hondius, a Fleming living in London, was a friend of the historian Richard Hakluyt and the Elizabethans who were promoting an English overseas empire. Hondius engraved the world map, illustrating Drake's circumnavigation (Plate 42), and most likely also engraved Boazio's maps of the West Indian voyage. Hondius's distinctive wave pattern, which he further developed and used in his early atlas maps, helps to confirm this supposition. The "Sea Cornye" at lower left is a trigger fish originally drawn by John White, the naturalist-governor of the Virginia colony at Roanoke. Boazio placed the cartouche at upper left, perhaps to cover uncertain geographical features, such as the persistent false Sea of Verrazzano and the source of the St. Lawrence River.

The West Indian voyage was sponsored by the Queen, members of her court, and London merchants. Elizabeth wanted to demonstrate further England's sea power after Drake's circumnavigation, particularly at a time when England and Spain appeared to be heading for war. The expedition might also enrich her treasury, deny the King of Spain needed revenue, and provide England more time to prepare for a major confrontation. The Queen's patronage is indicated on the map by her coat-of-arms in the compass rose. Drake's ship, *Elizabeth Bonaventure*, one of the royal contributions to the expedition, is illustrated at bottom battling an armed galley.

Twenty-five vessels made up Drake's fleet, each carrying about one hundred men, including twelve companies of soldiers. After stopping in northwest Spain and burning the town of Vigo, the expedition proceeded to Santiago in the Cape Verde Islands and took this outpost in November 1585. Although Santiago was very poor, yielding little plunder, the English troops burned it to the ground. Continuing to Santo Domingo, Hispaniola, Spain's first capital in the New World, Drake and his men discovered it in a state of decline, even though it remained one of the largest Spanish settlements in America. They captured the town with little trouble, but had overestimated its wealth and were obliged to settle for a modest ransom before angrily departing. One week later Drake's forces seized a weakly defended Cartagena, the most important city in the Caribbean. The English remained two months, suffering serious illness among crew and troops, then left with 100,000 ducats and 200 slaves.

The original plan called for storming Havana next, but they diverted to the Florida coast and burned St. Augustine before continuing on to Roanoke, Raleigh's English colony inside the Carolina banks. Drake offered the colonists, including John White (Plates 43 and 44) and the scientist-chronicler of the colony, Thomas Harriot, passage home. They accepted enthusiastically and reached Portsmouth in July 1586. The success of Drake's enterprise brought the investors a seventy-five percent profit.

Results of this successful West Indian voyage were two-fold: war between England and Spain became inevitable and reached its climax with the battle of the Spanish Armada in the English Channel in 1558; and Drake's name and fortune grew to legendary proportions.

References: British Museum 1977, no. 116; 1940, 58-60; Brown (comp.) 1952, no. 230; Church 1907, no. 136; Cumming 1962, 10; Cumming, Skelton & Quinn 1972, no. 314; JCB 1937, 12-14; 1940, 58-60; Keeler (ed.) 1981; Kraus 1970, 119-30; Waters 1955, 53-70; Winsor 1889, 3:82.

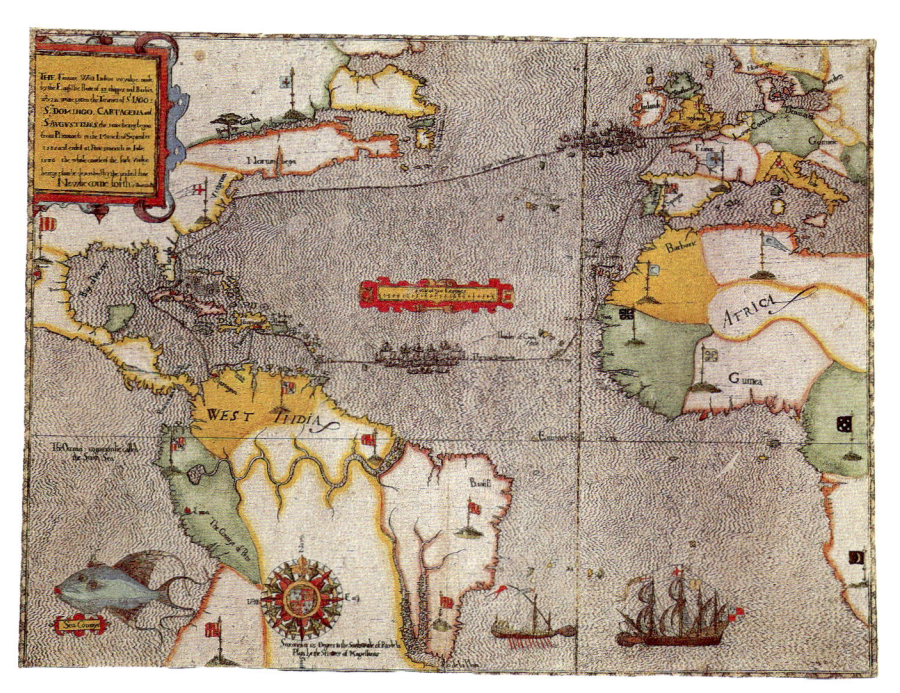

PLATE 45
Baptista Boazio (fl. 1588–1606)
The Famouse West Indian Voyadge...
Leiden (Netherlands), 1588

Copperplate map on paper, colored by hand, 16 x 21 in. (400 x 525 mm.)
Location:
Paul Mellon Collection, Upperville, VA

PLATE 46
Baptista Boazio (fl. 1588–1606)
Civitas S. Dominici sita in Hispaniola
Leiden (Netherlands), 1588

Copperplate engraving, colored by hand, 16 x 21 in. (400 x 550 mm.)
Location:
National Maritime Museum, Greenwich

PLATE 47
Baptista Boazio (fl. 1588–1606)
Civitas Carthagena in Indiae Occidentalis
Leiden (Netherlands), 1588
Copperplate engraving, colored by hand, 16 x 22 in. (400 x 550 mm.)
Location:
National Maritime Museum, Greenwich

PLATE 48
Baptista Boazio (fl. 1588–1606)
S. Augustini pars est terrae Florida
Leiden (Netherlands), 1588

Copperplate engraving, colored by hand, 18 x 22 in. (460 x 550 mm.)
Location:
Paul Mellon Collection, Upperville, VA

THE Famouse West Indian voyadge made by the Englishe fleete of 23 shippes and Barkes wherin weare gotten the Townes of S: IAGO: S: DOMINGO, CARTAGENA and S: AVGVSTINES the same beinge begon from Plimmowth in the Moneth of September 1585 and ended at Portesmouth in Iulie 1586 the whole course of the saide Viadge beinge plainlie described by the pricked line Newlie come forth by Baptista B.

PLATE 45
Baptista Boazio
The Famouse West
Indian Voyadge...

BAPTISTA BOAZIO, *HISPANIOLA*, LEIDEN, 1588

Santo Domingo was founded as the capital of the first Spanish colony in the New World by Christopher Columbus's brother, Bartolommeo, in 1496. It is the oldest permanent city in America established by Europeans. Originally named Nueva Isabella after the Queen, it was destroyed by hurricane, rebuilt, and renamed in 1502. Santo Domingo was the base from which the Spanish took control of other West Indian Islands and the nearby Spanish Main. After the conquest of Mexico and Peru, its importance declined; but the city remained populous and, so the English thought, prosperous.

Santo Domingo was Drake's first target on the West Indian raid. He arrived New Year's Day, 1586, having captured a small Spanish vessel nearby and forced its pilot to guide the English fleet to a safe landing ten miles west of the harbor. The soldiers, under Lt. Gen. Christopher Carleill, quickly captured the poorly defended city, which Drake proceeded ruthlessly to sack. He demanded a ransom for the city of one million ducats. To his dismay, the impoverished Spanish colonists could pay only 25,000.

Boazio's plan shows at left the guarded landing point and Carleill's English troops outside the city. The gardens at right, across the Ozama River, are on the site of the original settlement. In the city, the cathedral dominates the scene, while the governor's palace, once Bartolommeo Columbus's home, is on the quay. Decoration includes new fauna for a European audience: an alligator, a flying fish, a giant sea-turtle, and, surmounting the windrose, a winged mermaid indicating the north orientation.

References: British Museum 1977, no. 117; Keeler (ed.) 1981; Kraus 1970, 119-30; Waters 1955, 53-72.

BAPTISTA BOAZIO, *CARTAGENA*, LEIDEN, 1588

Drake arrived with his fleet in February 1586 at Cartagena, the most important city in the Caribbean. It served as a gathering point to which coastal freighters brought gold, pearls, hides, and other products of the Spanish Main. The annual convoy from Spain made its first stop here to pick up treasure, then sailed to Panama for Peruvian gold and other riches. Mexico and Cuba were the next ports of call before the fleet returned to Andalusia.

Cartagena, founded in 1533, became not only the chief entrepôt for northern South America but also a notorious center for the inquisition and the slave trade. In 1586 when news circulated that Drake, scourge of the Spanish, was in the region, the city should have heavily defended its natural harbor. Instead, as in Santo Domingo, Drake's forces made quick work of its capture. The now-familiar pillaging took place, and the English burned several buildings before a ransom of 100,000 ducats saved the remainder of the city from the torch.

Illness struck Drake's soldiers and crew at Cartagena, and the expedition was forced to remain for two months, to the acute displeasure of the local citizenry. During this period the British heard rumors that the Spanish were gathering an armada to strike back and would be waiting for Drake's fleet at Havana. Drake changed plans, and leaving Cartagena, avoided the enemy fleet by sailing west of Cuba and heading up the Bahamas channel to St. Augustine.

The capitulation of the fortified Spanish strongholds to the English, although disastrous, did not cause their total collapse. Shipments of treasure from Mexico, Peru, the Spanish Main, and West Indian Islands, and even from the Philippines, resumed. But the blow struck by Drake had calamitous short- and long-term results for Spain. Spanish credit crashed and morale deteriorated. The banks in Spain were destroyed and Philip II's foreign bankers in Italy and Germany withdrew their support. The long-range result was an increased British confidence and momentum that led to the founding of permanent colonies in Virginia, New England, and maritime Canada, eventually making North America (excluding Mexico and Canada) essentially an English domain.

References: British Museum 1977, no. 118; Keeler (ed.) 1981; Kraus 1970, 119-30; Waters 1955, 53-72.

BAPTISTA BOAZIO, *ST. AUGUSTINE*, LEIDEN, 1588

Boazio's rendition of St. Augustine, oldest city in the United States, was drawn from sketches brought back by Sir Francis Drake. It is the first published map of any United States city.

Ponce de León, searching for the legendary fountain of youth, claimed Florida for Spain in 1513. While unauthorized voyages could have visited this coast as early as 1500, no lasting settlement was made until 1565, when Philip II appointed Pedro Menéndez de Avilés Governor and Captain-General of Florida. Menéndez, upon learning that French Huguenots had built Ft. Caroline near the mouth of the St. Johns River the previous year, immediately led an expedition to destroy the settlement. French soldiers and settlers not killed in the attack were reported to have been hanged the next day under a sign that read, "I do this not to Frenchmen, but to Heretics." Menéndez then founded St. Augustine, twenty-five miles to the south. This outpost enabled Spain to assert its sovereignty over Florida, protect shipping lanes for the return voyage of her convoys, and plant the seeds of a North American colony.

The massacres of Spaniards by Frenchmen and Frenchmen by Spaniards were precipitated, like so much of the hostility of the era, by religious as well as national fervor. Two years after the Menéndez massacre, the French under the corsair De Gourgues took their revenge on the Spanish at the rebuilt Ft. Caroline, then called San Mateo. The fort was again destroyed, the Spanish garrison put to the sword, and the prisoners executed. This time an inscription was set up at the hanging that read, "Not as Spaniards, but as Traitors, Robbers, and Murderers."

Menéndez's slaughter of the Huguenots was on Drake's mind when he stopped at St. Augustine in 1586 en route from Cartagena to Roanoke. The Spanish had hastily built a fort at the seaward entrance to protect the city. Drake found it undefended and destroyed it. He then advanced on the town and is said to have taken particular relish in burning every structure to the ground. He was able to reprovision his fleet and refresh his men at St. Augustine before sailing on to learn how his countrymen were faring in Sir Walter Raleigh's Virginia colony at Roanoke.

The plan by Boazio, oriented with west at the top, shows the outer banks in the foreground with Drake's main fleet anchored off shore. Part of his army is deployed along the south side of the inlet, the rest is sacking the town at upper left, from where they carried away 2000 pounds of gold bullion. Boazio, who had accompanied Drake on the voyage as artist and interpreter, has left us an accurate picture of this historic episode.

References: British Museum 1977, no. 119; Brown (comp.) 1952, no. 230; Cumming, Skelton & Quinn 1972, no. 224; Keeler (ed.) 1981; Kraus 1970, no. 49; Waters 1955, 53-72.

PLATE 46
Baptista Boazio
Civitas S. Dominici
sita in Hispaniola

NEC SPE NEC METV

CIVITAS CARTHAGENA in Indiae occidentalis continente sita, portu commodissimo, ad mercaturam inter Hispaniam et Peru exercendam

CARTAGENA

PLATE 47
Baptista Boazio
Civitas Carthagena in Indiae Occidentalis

OPIDVM S. Augustini lignies ædibus constructum, amœnissimos habuit hortos, eóq; solo fœcundissimo à nobis vero cum inde solverimus incensis una in cineres redactum. Præsidium hic erat 150 Hispanorum, aliudq; item eodem numero ad duodecim Septentrionem versus leucas in loco S. Helenæ dicto, hæc cum præ- sidia quemadmodum etiam in præsenti non alio consilio disposita erant nisi ad prohibendos Anglos et Gallos ne intersectum regionem quin prorsus ventri græci, occuparent.

PLATE 48
Baptista Boazio
S. Augustini pars est terrae Florida

CORNELIS DE JODE, *QUIVIRIAE REGNUM* & *AMERICAE PARS BOREALIS*, ANTWERP, 1593

Almost all early maps of America encompass the entire Western Hemisphere; however, these two engravings make up the most advanced map published to that date devoted solely to North America. In the Low Countries, leading center of cartographic publishing, De Jode was the only mapmaker to express his belief in the importance of North America by compiling a large three-page map devoted exclusively to that continent. This is the first general production to use the eyewitness delineations of Jacques Le Moyne's *Florida* and John White's *Virginia* (Plates 43 and 44). It represents the point of departure from which John Smith in his *Chesapeake Bay* and *New England*, and Samuel de Champlain in his *Canada* began their pioneer explorations and charting in the following decade.

Inscribed throughout are references to important events such as the unsuccessful colonization attempts by French Huguenots in Carolina and Florida and by Raleigh's Elizabethans in Virginia (North Carolina). Further west, fabled names appear that had fascinated Europeans since the Middle Ages: the Seven Cities, Cibola, Anian, and Quivira.

By the time of Columbus, sea charts reflected returning ocean voyagers' reports of sighted islands, real and imaginary. Whether provoked by their own superstitions or by violent storms at sea where breaking waves strongly resembled the surf on a shoreline, these men were convinced they had seen land. Mapmakers, always averse to blank areas on their charts, were pleased to make additions based on such reports. Features thus added tend to remain, even after proven false.

As Europe probed the Ocean Sea westward, beginning with Columbus and his contemporaries, the location of these legendary places receded further westward. De Jode's map shows them at their ultimate retreat, Anian relocated in what will become Alaska, and Quivira posited in California. The Seven Cities, originally reported on Antilia, a mythical island in the Atlantic in medieval times, have now migrated to New Mexico. There, as De Jode notes, the Franciscan missionary and explorer Friar Marcos de Niza thought he discovered the Cities upon sighting the Zuñi pueblos. Niza's reports in turn prompted the Coronado expedition to the Southwest that opened that area to Spanish settlement.

Cornelis De Jode had succeeded to the map and atlas publishing business of his father, Gerard (1515–91). Fifteen years earlier the elder De Jode had published an atlas which at that time was the only competitor to the work of Abraham Ortelius. Cornelis updated, corrected, and expanded this second edition; and *North America* and *Quivira* are two of the important new maps he prepared.

In the Pacific off California, there are two fine European ships and a large oriental sailing galley. They share that stretch of ocean with two giant sea monsters: one is half unicorn, half fish, the other, a spiny-backed dragon perhaps inspired by a marine iguana. The panel in the Atlantic depicts six full-length portraits of Virginia Indians originally drawn by John White, the naturalist of the first Virginia colony. The vignette at top right shows Eskimos attacking Martin Frobisher's ship with bows and arrows, depicting an incident from an earlier voyage when White had accompanied Frobisher to the Arctic.

Cornelis De Jode focused more attention on the North American continent than did any of his contemporaries. This compilation, using current sources and containing considerable detail and many place-names, suffers only from cartographic problems that could not be solved in De Jode's time. Determination of longitude was still imprecise, knowledge of the continental interior and Arctic region was insufficient, and the high latitudes were then, as now, difficult to portray except on polar projection maps, which introduced still other problems.

In 1593, when the sixteenth century and the first century of European overseas discovery were coming to a close, De Jode presented the sum of current geographical knowledge of North America.

References: Cumming 1962, no. 16; Cumming, Skelton & Quinn 1972, no. 242; Heidenreich & Dahl 1980; Koeman 1969, Jod 2:12 & 13.

PLATE 49A
Cornelis De Jode (1568–1600)
Quiviriae Regnum
Antwerp, 1593
Copperplate engraving, colored by hand, 13.5 x 9 in. (343 x 229 mm.)
Location:
The Newberry Library, Chicago

QVIVIRÆ REGNV,
cum alijs versus Boreã.

Septentrio.

Oceanus 19 ostijs inter has insulas irrumpens, 4 euripos facit, quibus indesinenter in Septentrionem fertur, atq́ ibidem, mirè vehementerq́ absorbetur.

Polus Magnetis respectu insularū Capitis Viridis.

El Streto de Anian.

Circulus Anian Reg.

Bergi.

Hic hominū societates cernuntur ruri, in tentorijs habitantes, more Hordarū, quas apud Tartaros videmus.

Pagul

Regio hæc plana est et silvestris, in qua boves, vaccæq́ reperiuntur, gibbū camelorū habentes, cauda vero, et pedibus leones referūt.

C. Blanco
Tierra frisida
B. Hermosa
C. de Fortuna
C. de Corrientes
Tierra medicina
Tuchano
C. Mendocino
Cabo de Corrientes
Cabo Mendocino
B. de Trabaio
B. Hermosa
Plaia
Quivira Regnū.
Quivira
R. Hermosa
C. de S. Francisco
Plaia
B. de S. an
Tierra delos Pescadores
C. Blanco
Plaia
R. Grande
R. Hermosa
Los Frailes
La sierra nevada
C. Blanco

Las dos Hermanas

Las Mongos
La Vezina
La desgraciada

OCEANVS

Meridies.

Generoso, atq; Magnifico Dno, Dno THEODORICO ECHTER, à Meſpelbrū, Sacr. Cæſar. Maieſt.ⁿ et Reverˢᵐᵒ Principi, Epiſcopo Herbipolenſi a conſilijs primo &c Cornelius de Iudæis Antverp. D. D. Aº MD LXXXXIII.

AMERICÆ REALIS, FLOR OS, CANADA LIS. A Cornelio de

Regiones hæ multū adhuc ſunt incognitæ, neq; eo ob intēſiſimum frigus, adnavigare licet.

IN COG

Perpetuis nivibus hæ Regiones coopertæ perhibentur.

Arcticus.

Incolæ harū Regionū, piſcibus magna ex parte vivunt, et pellibus ferarū veſtiuntur.

Saguen

Hoc fluvio facilis navigatio eſt in Saguenai

De ſtatu Regionis Virginiæ.
Eam Regionem, quā hodie Virginiam appellamus, A.º 1585. ſumptibus Walteri Raleigh, Angli detexerūt. Regio, Aluminis, Vini, ferri, Æris, Cedri, omniū neceſſariorū feraciſſima: frumēti incredibilis prouētus. Gens mediocri ſtatura, chlamyde è ceruina pelle, tecta Arma, arcus ſunt, et ſtipites lignei, oppida exigua, mari vicina ſunt, ſo aut 12 ædes, rarò apſius, habēt. Pœna etia in delinquētes inſtitutæ ſunt.

Egis flu mare dulcium aquarum eſt, cuius terminū ignorari Canadenſes aiunt.

Chiggigui Ohio flu Canaoga Combas Lago de Conibas

Salboy Cubirao Subilaga

In his montibus habitāt diverſæ nationes, homines feri et ſine lege; quiq; continuis bellis inter ſe conflictantur, Auanares ſcil. Albardi, Calicuaz, Tagil, Apalche, Chilaga, Mocoſa, pluresq; aliæ.

Tolgago Zubgara Hochalaga Hochalag

Qui inter Floridam et terram Baccalaos habitāt, hi omnes uno nomine Canadenſes appellantur: ſed diuerſæ nationes populorum, ut ſunt Hochelaga, Honguedo, Corterealis, præ cæteris benigni et humani.

Chichuco Septem Citta Marcus Niza aſſeverat, Prouinciam Septē Cittā valde eſſe nobilem. Albardos Auanares NOVA

Axa Chuce Ceuola Naguater Capaſchi Tagil

Tontonteac Abacus flu. Granada Nicoſa Choque Laconia Chillano Aux Nualatino Tuſcia Gayar Achus FLORIDA

CALIFORNIA Suala Mons Marata Artula Terchichi mechi Coſſa

Tiguex Farillones Tierras pl. P. de S. Ma P. de S. Iago I. de Salamary P. de la Cruz Golfo Bermejo Guianal flu. Ometlan Cecos Culia Chichaca Withkoholi

Cabo del Engano Ca̧zones I. de Cedros Los dandros Sebast Cualis cana Toua Golfo de Mexico

Tropicus Cancri I. de Cedros Tortuga Havana

PLATE 49B
Cornelis De Jode (1568–1600)
Americae Pars Borealis
Antwerp, 1593
Copperplate engraving, colored by hand,
14.5 x 20 in. (368 x 508 mm.)
Location:
The Newberry Library,
Chicago

EDWARD WRIGHT, WORLD CHART ON MERCATOR PROJECTION, LONDON, 1599

The sixteenth century ended with Elizabethan advocates of a English empire gaining the support of Queen and court. These were bold men: great navigators, seamen, explorers, pirates, (sometime slavers), and the nemeses of Spanish outposts; men such as John and Richard Hawkins, Francis Drake, Martin Frobisher, and John Davis. Others were colonial promoters, Humphrey Gilbert, Walter Raleigh and Richard Grenville; intellectuals and propagandists, John Dee and Richard Hakluyt; scientists, William Gilbert and Edward Wright; and many more. Among them they maneuvered as effectively at home as at sea to set the stage for establishing English presence in America.

The first Virginia colony, Raleigh's Roanoke inside the Carolina banks, ultimately was abandoned, in part because of Elizabeth's domestic problems and the war with Spain. However, the efforts of Ralph Lane, first governor; John White, naturalist and later governor; and Thomas Harriot, scientist and chronicler of the enterprise, paved the way for Jamestown on Chesapeake Bay and eventually Plymouth, Massachusetts.

If there were a single book that personified the coming of age of Elizabethan England, it was Richard Hakluyt's *The Principal Navigations, Voiages, Traffiques and Discoveries of the English Nation*. Published in London, 1598–1600, in three folio volumes, it contains over 1.5 million words and is the most thorough collection of British voyages and discoveries by land and sea in existence. It is aptly called the "prose epic of the English nation."

What might be considered the *graphic* epic of the English nation is this anonymous map, designed by Edward Wright and selected by Richard Hakluyt as the only illustration to augment his *Principal Navigations*. Wright, author of the scientific treatise, *Certaine Errors in Navigation*, drew upon a recently issued globe by Emery Molyneux (the first to be made in England) for his delineation. He projected it upon a Mercator grid of straight latitude and longitude lines intersecting at right angles. His friends Hakluyt and John Davis made certain his geographical information was correct and that the map accurately summarized current maritime knowledge.

As the cartouche below Africa explains, only discovered coasts are shown; for example, the unknown area north of Cape Mendocino, California, is completely blank. In contrast to most world maps of this period, including the Hondius double-hemisphere, tracking Drake's circumnavigation (Plate 42), there is no oversized Antarctic continent on Wright's map. The discoveries of Drake, Cavendish, and others recorded in Hakluyt's text are reflected, improving the map over its predecessors. "Queen's Island" at the bottom of South America refers to "Elizabetha," the name Drake put on Cape Horn when contrary winds blew him there on his circumnavigation. Very few places inland are identified, notable exceptions being the Amazon and Rio de La Plata systems.

In North America a complex and confusing hydrography seems to connect the St. Lawrence, the Hudson, Lake Tadouac (early representation of the Great Lakes on a map), and an unidentified coastline interpreted either as a receding Sea of Verrazzano or an incipient Hudson's Bay. Although previous maps had shown Greenland contiguous with Labrador, Davis Strait is named and correctly located between Greenland and Baffin Island. Lands north of the St. Lawrence are called Canada, while Virginia covers everything southward. Recent discoveries are indicated by the appearance of Chesapeake, Hatteras, and Cape Fear along the coastline. Place-names have been drawn not only from English and native American originals but also from Portuguese and French sources.

Famous in its own day, this is the "new map" referred to by Shakespeare in *Twelfth Night*, performed in 1601: "He does smile his face into more lines than is in the new map with the augmentation of the Indies."

The Wright-Molyneux-Mercator-Hakluyt chart represents the culmination of cartographic breakthroughs that began with Christopher Columbus and the first Spanish and Portuguese navigators. These explorers freed Europe from the constraints of the world as known to the ancients and brought Europeans knowledge of new worlds in the East and West Indies. Comparing this map with that of Ptolemy (Plate 1) dramatizes the astonishing advances in geographical knowledge achieved in one century.

References: Bagrow/Karrow 1990, no. 56:17.8; Brown (comp.) 1952, no. 166; Church 1907, no. 322; Cumming, Skelton & Quinn 1972, no. 269; Fite & Freeman 1926, no. 28; Ganong 1964, 456-61; Nordenskiöld [1889] 1961, 96; Parsons & Morris 1939, 61-71; Shirley 1983, no. 221; Wagner 1926, 422; 1937, no. 224; Wallis 1974, 69-73; Wroth 1944, no. 64.

PLATE 50 (detail). *Edward Wright, America*

DIEV ET MON DROIT

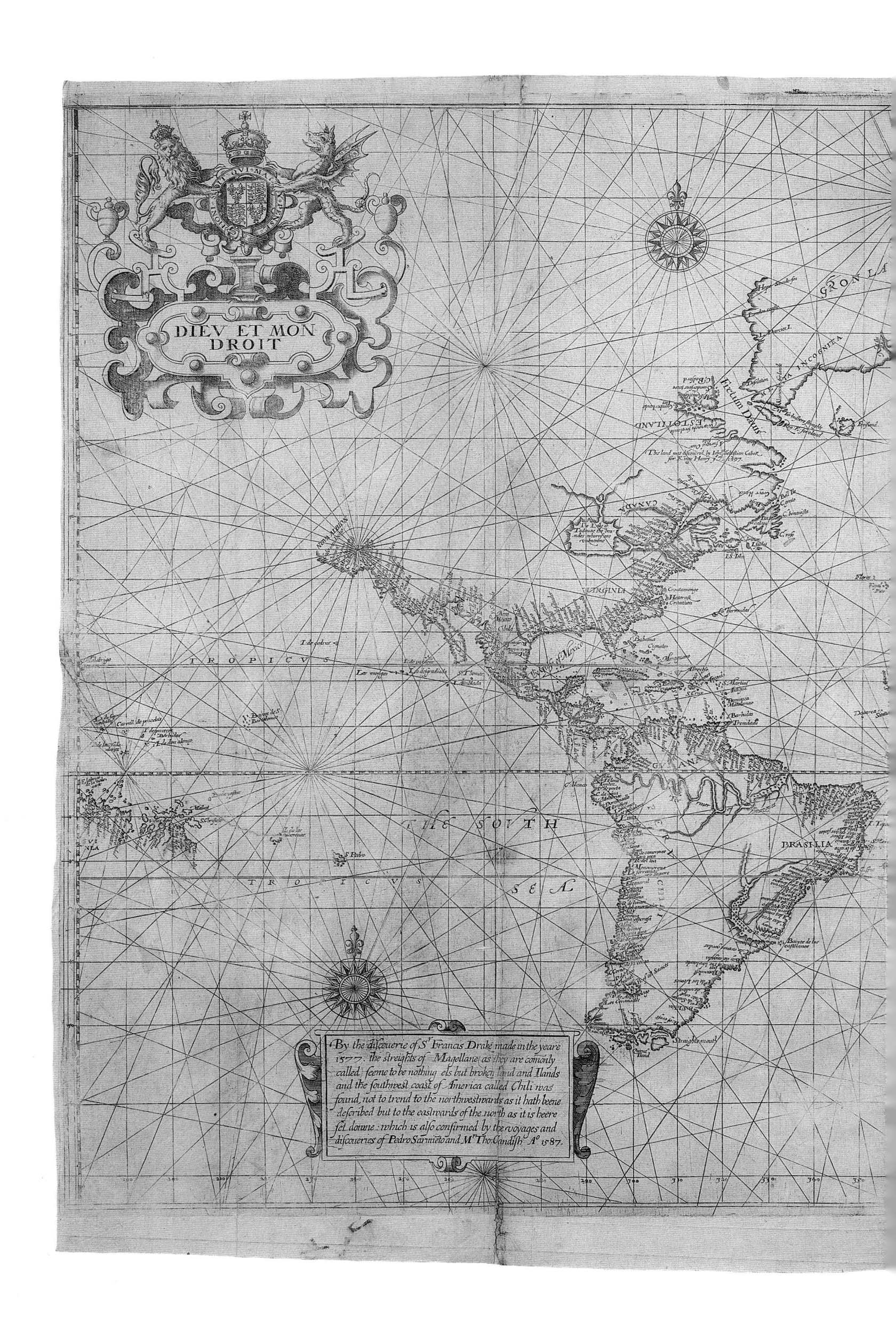

By the discouerie of S.r Francis Drake made in the yeare 1577. the streights of Magellane as they are comonly called, seeme to be nothing els but broken land and Ilands and the southwest coast of America called Chili was found, not to trend to the northwestwards as it hath beene described but to the eastwards of the north as it is heere set downe : which is also confirmed by the voyages and discoueries of Pedro Sarmieto and M.r Tho: Candish A.o 1587.

PLATE 50
Edward Wright (1558–1615)
World Chart on Mercator Projection
London, 1599
Copperplate engraving,
16.5 x 25 in. (355 x 755 mm.)
Location:
The Newberry Library,
Chicago

BIBLIOGRAPHY

Included are references cited in the text, sources used by the author, and recommended studies for further reading.

Adonias, Isa. 1963. *A Cartografia da Regiao Amazonica.* 2 vols. Rio de Janeiro.

Afetinan, A. 1954. *Life and Works of the Turkish Admiral: Piri Re'is,* trans. Leman Yolac. Ankara: Turk Tarih Kurumu Basimevi.

Akcura, Yusuf. 1935. *Piri Re'is' Map.* Istanbul: Devlet Basimevi.

Almagià, Roberto. 1944-55. *Monumenta Cartographica Vaticana.* 3 vols. Vatican City: Biblioteca Apostolica Vaticana.

———. 1948. The First "Modern" Map of Spain. *Imago Mundi.* 5:27-31.

———. 1956. *Il Mappamundo di Fra Mauro.* Venice: Istituto Poligrafico dello Stato.

Andrews, Kenneth R. 1984. *Trade, Plunder and Settlement. Maritime Enterprise and the Genesis of the British Empire, 1480-1630.* Cambridge: Cambridge University Press.

Babcock, William H. 1922. *Legendary Islands of the Atlantic: A Study in Medieval Geography.* New York.

Bagrow, Leo. 1964. *History of Cartography,* revised and enlarged by R.A. Skelton. Cambridge: Harvard University Press.

Bagrow, Leo and Robert Karrow. 1990. *Mapmakers of the Sixteenth Century and their Maps.* The Catalog of Cartographers of Abraham Ortelius, 1570. Chicago: Speculum Orbis Press and The Newberry Library (in preparation).

Barreiro-Meiro, Roberto. 1970. *Las Islas Bermudas y Juan Bermudez.* Madrid: Instituto Historico de Marina.

Beazley, Charles Raymond. 1897-1906. *The Dawn of Modern Geography: A History of Exploration and Geographical Science from the Conversion of the Roman Empire to A.D. 900.* 3 vols. London: J. Murray.

Bibliothèque Nationale. 1979. *A La Découverte de La Terre. Dix Siècles de Cartographie.* Paris: Bibliothèque Nationale.

Bigelow, John. 1935. The so-called Bartholomew Columbus Map of 1506. *Geographical Review.* 25: 643-56.

Biggar, Henry Percival. 1911. *The Precursors of Jacques Cartier.* Publications of the Public Archives of Canada, No. 5. Ottawa.

———. 1924. *The Voyages of Jacques Cartier.* Publications of the Public Archives of Canada, No. 11. Ottawa.

Boorstin, Daniel J. 1983. *The Discoverers.* New York: Random House.

Bradford, Ernle. 1973. *Christopher Columbus.* New York: Viking.

British Museum. 1977. *Sir Francis Drake, An Exhibition to Commemorate Francis Drake's Voyage around the World 1577-1580.* London: British Museum for the British Library.

Brown, Lloyd A. 1950. *The Story of Maps.* Boston: Little, Brown.

———, comp. 1952. *The World Encompassed.* Baltimore: Walters Art Gallery.

Bunbury, Edward Herbert. [1883] 1959. *A History of Ancient Geography among the Greeks and Romans from the Earliest Ages till the Fall of the Roman Empire.* 2d ed. 2 vols. London. Reprint, with new introduction by E.H. Stahl. New York: Dover.

Campbell, Tony. 1981. *Early Maps.* New York: Abbeville.

———. 1987. *The Earliest Printed Maps 1472-1500.* London: British Library.

Caraci, G. 1937. A Little Known Atlas by Vesconte Maggiolo, 1518. *Imago Mundi.* 2:37-54.

Church, Elihu Dwight. 1907. *A Catalogue of Books Relating to the Discovery and Early History of North and South America,* comp. G.W. Cole. 5 vols. New York.

Cortesão, Armando. 1954. *The Nautical Chart of 1424 and the Early Discovery and Cartographical Representation of America; a Study on the History of Early Navigation and Cartography.* Coimbra: University of Coimbra.

———. 1969-71. *History of Portuguese Cartography.* 2 vols. Coimbra: Junta de Investigações do Ultramar-Lisboa.

Cortesão, Armando and A. Teixeira da Mota. 1960. *Portugaliae Monumenta Cartographica.* 6 vols. Lisbon.

Crone, Gerald R. 1961. Martin Behaim, navigator and cosmographer, figment of imagination or historical personage? *Congresso Internacional de Historia dos Descobrimentos, Actas.* 2:117-33.

———. 1968. *Maps and Their Makers. An Introduction to the History of Geography.* London: Hutchinson.

Cumming, William P. 1962. *The Southeast in Early Maps.* Chapel Hill: University of North Carolina Press.

———. 1988. *Mapping the North Carolina Coast—Sixteenth Century Cartography and the Roanoke Voyages.* Roanoke: North Carolina Department of Cultural Resources.

Cumming, William P., R.A. Skelton, and David B. Quinn. 1972. *The Discovery of North America.* New York: American Heritage.

Davies, Arthur. 1977. Behaim, Martellus and Columbus. *Geographical Journal.* 143:451-9.

De Costa, B.F. 1881. *Cabo de Baros or the Place of Cape Cod in the old Cartology.* New York: Whittaker.

Destombes, Marcel. 1964. *Mappemondes A.D. 1200-1500.* Amsterdam: N. Israel.

———. 1987. *Contributions selectionnées a l'Histoire de la Cartographie et des Instruments Scientifiques....* Utrecht: HES.

Fischer, Joseph and R. von Wieser. 1903. *The Oldest Map with the Name America of the Year 1507 and the Carta Marina of the Year 1516 by M. Waldseemüller.* Innsbruck.

Fite, Emerson D. and Archibald Freeman. 1926. *A Book of Old Maps Delineating American History from the Earliest Days down to the Close of the Revolutionary War.* Cambridge: Harvard University Press.

Fuson, Robert H., trans. 1987. *The Log of Christopher Columbus.* Camden, Me.: International Marine Publishing Co.

Ganong, W.F. 1964. *Crucial Maps in the Early Cartography and Place-Nomenclature of the Atlantic Coast of Canada,* ed. T.E. Layng. Toronto: University of Toronto Press.

George, Wilma. 1969. *Animals and Maps.* Berkeley: University of California Press.

Gerace, Donald T., comp. 1986. *First San Salvador Conference—Columbus and His World. Proceedings.* San Salvador Is., Bahamas: College Center of Finger Lakes.

Giraldi, Alberto, ed. 1954-55. *Collection of Maps and Documents Shown at the Exhibition Held in the Palazzo Vecchio in Florence on the Quincentenary of the Birth of Amerigo Vespucci.* Florence.

Goldstein, Thomas. 1965. Geography in Fifteenth-Century Florence. In *Merchants & Scholars,* ed. John Parker. Minneapolis: University of Minnesota Press.

Grosjean, George, ed. 1978. *Mapamundi. The Catalan Atlas of the Year 1375.* Dietikon-Zurich: Urs Graf.

Harley, J.B. and David Woodward. 1987. *The History of Cartography.* Vol. 1, *Cartography in Prehistoric, Ancient, and Medieval Europe and the Mediterranean.* Chicago and London: University of Chicago Press.

Harrisse, Henry. 1866. *Bibliotheca Americana Vetustissima. A Description of Works Relating to America, published between the years 1492 and 1551.* New York.

———. 1892. *The Discovery of North America.* London and Paris.

———. 1900. *Découverte et Evolution Cartographique de Terre Neuve et les Pays Circonvoisins, 1497-1501-1769.* London and Paris.

Heidenreich, Conrad E. and Edward H. Dahl. 1980. The French Mapping of North America in the Seventeenth Century. *The Map Collector.* 13 (Dec.):2-11.

Hoffman, Bernard G. 1961. *Cabot to Cartier. Sources for a Historical Ethnography of Northeastern North America 1497-1550.* Toronto: University of Toronto Press.

Hough, Samuel J. 1980. *The Italians and the Creation of America. An Exhibition at the John Carter Brown Library.* Providence: Brown University.

Howse, Derek and Michael Sanderson. 1973. *The Sea Chart. An Historical Survey based on the Collections in the National Maritime Museum.* New York: McGraw-Hill.

Hulton, Paul. 1984. *America 1585—The Complete Drawings of John White.* Chapel Hill: University of North Carolina Press.

Hulton, Paul and David B. Quinn, eds. 1964. *The American Drawings of John White 1577-1590.* 2 vols. London and Chapel Hill.

Imago Mundi—The Journal of the International Society for the History of Cartography. 1935-1989. Vols. 1-41, continuing. Berlin and London.

Jantz, Harold. 1976. The New World in the Treasures of an Old European Library. *Exhibition of the Duke August Library.* Wolfenbüttel.

[J.C.B.] *Bibliotheca Americana. Catalogue of the John Carter Brown Library in Brown University.* 1919-31. 3 vols. Providence.

John Carter Brown Library. 1901-1972. Brown University Annual Reports 1901-1966. Providence.

Judge, Joseph, et al. 1986. Where Columbus Found the New World. *National Geographic.* 170 (no. 5):566-72, 578-99.

Kahle, Paul. 1933a. *Die Verschollene Columbus-Karte von 1498 in Einer Türkischen Weltkarte von 1513.* Berlin: Walter de Gruyter.

———. 1933b. A Lost Map of Columbus. *Geographical Review.* 23:622-3.

Keeler, Mary Frear, ed. 1981. *Sir Francis Drake's West Indian Voyage.* London: The Hakluyt Society.

Kelsey, Harry. 1987. The Planispheres of Sebastian Cabot and Sancho Gutierrez. *Terrae Incognitae.* 19:41-58.

Kimble, George H.T. 1938. *Geography in the Middle Ages.* London: Methuen.

Kish, George. 1965. The cosmographic heart: cordiform maps of the 16th century. *Imago Mundi.* 19:13-21.

———. 1966. Two Fifteenth-Century Maps of "Zipangu": Notes on the Early Cartography of Japan. *Yale University Library Gazette.* 40(no. 4): 206-14.

Koeman, C., comp. 1967-71. *Atlantes Neerlandici.* 5 vols. Amsterdam: Theatrum Orbis Terrarum.

Kohl, J.G. 1869. *History of the Discovery of Maine.* Publications of the Maine Historical Society, vol. 1. Portland.

Kraus, Hans P. 1970. *Sir Francis Drake. A Pictorial Biography.* Amsterdam: N. Israel.

Kunstmann, Friedrich. 1859. *Atlas zur Entdeckungsgeschichte Amerikas.* Munich.

La Roncière, Charles de. 1924. *La Carte de Christophe Colomb.* Paris.

Lach, Donald F. 1965. *Asia in the Making of Europe.* Vol. 1, The Century of Discovery, Book One and Book Two. 2 vols. Chicago: University of Chicago Press.

Layng, T.E., comp. 1956. *Sixteenth-Century Maps Relating to Canada. A Check-list and Bibliography.* Ottawa: Public Archives of Canada.

———. 1961. Charting the course to Canada. In *Congresso Internacional de Historia dos Descobrimentos, Actas.* 2: 255-76.

Lowery, Woodbury. 1912. *The Lowery Collection. A Descriptive List of Maps of the Spanish Possessions within the present limits of the United States, 1502-1820,* ed. Philip Lee Phillips. Washington.

Major, Richard Henry. 1868. *The Life of Prince Henry of Portugal, Surnamed the Navigator.* London: Asher.

The Map Collector. 1977-89. Nos. 1-49 continuing. Tring, England.

Mapes, Carl. 1963. Accompanying text of Rand McNally & Company 1963 Christmas card showing the world map of Vesconte Maggiolo from his Portolan Atlas of 1511.

Marzoli, Carla Clivio, Giacomo Corna Pellegrini, and Gaetano Ferro. 1985. *Imago et Mensura Mundi. Atti del IX Congresso Internazionale di Storia della Cartografia.* 3 vols. Rome: Enciclopedia Italiana.

McCorkle, Barbara B. 1985. *America Emergent. An Exhibition of Maps and Atlases in Honor of Alexander O. Vietor.* New Haven: Yale University Press.

McGuirk, Donald L., Jr. 1986. The Mystery of Cuba on the Ruysch Map. *The Map Collector.* 36 (September):40-1.

———. 1989a. The Depiction of Cuba on the Ruysch World Map. *Terrae Incognitae.* 20: 89-97.

———. 1989b. Ruysch World Map: Census and Commentary. *Imago Mundi.* 41: 133-41.

Mollat du Jourdin, Michel et al. 1984. *Sea Charts of the Early Explorers. 13th to 17th Century.* New York: Thames & Hudson.

Morison, Samuel Eliot. 1942. *Admiral of the Ocean Sea. A Life of Christopher Columbus.* 2 vols. Boston: Little, Brown.

———. 1971. *The European Discovery of America. The Northern Voyages A.D. 500-1600.* New York: Oxford University Press.

———. 1974. *The European Discovery of America. The Southern Voyages A.D. 1492-1916.* New York: Oxford University Press.

Nebenzahl, Kenneth. 1982. Accompanying text of Rand McNally & Company 1982 Christmas card showing the East Coast of North America by an anonymous chartmaker of Dieppe, France, circa 1547.

———. 1986. *Maps of the Holy Land.* New York: Abbeville.

Newton, R.R. 1977. *The Crime of Claudius Ptolemy.* Baltimore: Johns Hopkins University Press.

Nordenskiöld, A.E. [1889] 1961. *Facsimile-Atlas to the Early History of Cartography.* Stockholm. Reprint. New York: Kraus Reprint Corp.

———. [1897] n.d. *Periplus, An Essay on the Early History of Charts and Sailing-Directions.* Stockholm. Reprint. Burt Franklin, ser. 52. New York.

Nunn, George E. 1924. The Geographical Conceptions of Columbus. In *American Geographical Society,* Research Series No. 14. New York.

———. 1928. *World Map of Francesco Rosselli.* Philadelphia: Privately printed.

———. 1946. *The La Cosa Map and the Cabot Voyages. Was New York Bay Discovered by John Cabot, 1498?* Tall Tree Library, no. 19. Jenkintown, Pa.

———. 1948. *The Diplomacy Concerning the Discovery of America.* Tall Tree Library, no. 20. Jenkintown, Pa.

———. 1952. The Three Maplets Attributed to Bartholomew Columbus. *Imago Mundi.* 9:12-22.

Osley, A.S. 1969. *Mercator.* London: Faber & Faber.

Parker, John. 1955. *Antilia and America.* Minneapolis: James Ford Bell Collection, University of Minnesota Press.

———. 1965a. *Books to Build an Empire. A Bibliographical History of English Overseas Interests to 1620.* Amsterdam: N. Israel.

———, ed. 1965b. *Merchants and Scholars: Essays on the History of Exploration and Trade.* Minneapolis: University of Minnesota Press.

Parry, J.H. 1963. *The Age of Reconnaissance Discovery, Exploration, and Settlement 1450-1650.* Cleveland: World.

Parsons, E.J.S. and W.F. Morris. 1939. Edward Wright and His Work. *Imago Mundi.* 3:61-71.

Penrose, Boies. 1955. *Travel and Discovery in the Renaissance 1420-1620.* Cambridge: Harvard University Press.

Pohl, Frederick Julius. 1966. *Amerigo Vespucci, Pilot Major.* New York: Octagon Books.

Polaschek, Erich. 1959. Ptolemy's *Geography* in a New Light. *Imago Mundi.* 14:17-37.

Power, Robert H. 1973. Drake's Landing in California: A Case For San Francisco Bay. *California Historical Quarterly.* 52:100-30.

Putman, Robert. 1983. *Early Sea Charts.* New York: Abbeville.

Quinn, David B. 1955. *The Roanoke Voyages 1584-1590.* 2 vols. The Hakluyt Society, n.s., vol. 104. London.

———. 1977. *North America from Earliest Discovery to First Settlements. The Norse Voyages to 1612.* New York: Harper.

———. 1987. Bermuda in the Columbian Era. *Bulletin of the Institute of Maritime History and Archaeology.* 10:4-7, 19-25.

Ravenstein, E.G. 1908. *Martin Behaim—His Life and His Globe.* London: Philip.

Rey Pastor, Julio and Ernesto Garcia Camarero. 1960. *La Cartografia Mallorquina.* Madrid: Instituto Luis Vives.

Rogers, Francis M. 1958. *The Obedience of a King of Portugal.* Minneapolis: University of Minnesota Press.

———. 1966. Europe Informed. An Exhibition of Early Books Which Acquainted Europe With the East. In *6th International Colloquium on Luso-Brazilian Studies.* Cambridge: Harvard University Press.

The Rosselli Oval Planisphere (1507-1508). 1982. San Francisco: John Howell Books.

Rowse, A.L. 1959. *The Elizabethans and America. The Trevelyan Lectures at Cambridge—1958.* New York: Harper.

Sabin, Joseph. 1868-1936. *Bibliotheca Americana: A Dictionary of Books Relating to America.* 29 vols. New York.

Schilder, Gunther. 1986-88. *Monumenta Cartographica Neerlandica.* 2 vols & 2 portfolios. Alphen, Netherlands: Canaletto.

Schwartz, Seymour I. and Ralph E. Ehrenberg. 1980. *The Mapping of America.* New York: Abrams.

Schwartz, Stuart B. 1986. *The Iberian Mediterranean and Atlantic Traditions in the Formation of Columbus as a Colonizer.* Minneapolis: University of Minnesota Press.

Shirley, Rodney W. 1983. *The Mapping of the World—Early printed world maps 1472-1700.* London: Holland Press.

Skelton, R.A. 1958. *Explorers' Maps.* New York: Praeger.

———. 1963. Introduction to *Claudius Ptolemaeus Cosmographia, Ulm 1482.* Theatrum Orbis Terrarum, 1st ser., vol. 2. Amsterdam.

———. 1965a. Introduction to *Braun and Hogenberg: Civitates orbis terrarum (Cologne and Antwerp, 1572-1618).* Mirror of the World, 1st ser., vol. 1. Amsterdam: Theatrum Orbis Terrarum.

———. 1965b. *Looking at an Early Map.* Lawrence, Ks.: University of Kansas Library.

———. 1966. Introduction to *Claudius Ptolemaeus Geographia Strassburg, 1513.* Theatrum Orbis Terrarum, 2nd ser., vol. 4. Amsterdam.

———, trans. and ed. 1969. *Magellan's Voyage. A Narrative Account of the First Circumnavigation by Antonio Pigafetta.* New Haven and London: Yale University Press.

———. 1972. *Maps. A Historical Survey of Their Study and Collecting.* Chicago: University of Chicago Press.

Skelton, R.A., T.E. Marston, and G.D. Painter. 1965. *The Vinland Map and the Tartar Relation.* New Haven: Yale University Press.

Sprent, F.P. 1926. *A Map of the World Designed by Giovanni Matteo Contarini. Engraved by Francesco Rosselli. 1506.* 2nd ed. rev. London: British Museum.

———. 1927. *Sir Francis Drake's Voyage Round the World, 1577-1580. Two Contemporary Maps.* London: British Museum.

Stevens, Henry N. 1908. *Ptolemy's Geography. A Brief Account of all the Printed Editions Down to 1730.* 2nd ed. London: Stevens.

———. 1928. *The First Delineation of the New World and the First Use of the name America on a Printed Map.* London: Stevens.

Stevenson, Edward Luther. 1903. *Maps Illustrating Early Discovery and Exploration in America 1502-1520.* New Brunswick, N.J.

———. 1908. *Marine World Chart of Nicolo de Canerio Januensis 1502 (circa).* New York.

———. 1909. *Early Spanish Cartography of the New World.* Worcester, Ma.: Davis Press.

———. 1911. *Portolan Charts. Their Origin and Characteristics.* Publications of the Hispanic Society of America, no. 82. New York.

———. 1921. *Terrestrial and Celestial Globes.* 2 vols. Publications of the Hispanic Society of America, no. 86. New Haven.

Stokes, I.N. Phelps. 1915-28. *Iconography of Manhattan Island—1498-1909.* 6 vols. New York: Dodd.

Swan, Bradford. 1951. The Ruysch Map of the World (1507-1508). *Papers of the Bibliographical Society of America.* 45:219-36.

Sykes, Percy. 1950. *A History of Exploration. From the Earliest Times to the Present Day.* London: Routledge & Kegan Paul.

Taylor, E.G.R. 1957. *The Haven-Finding Art. A History of Navigation from Odysseus to Captain Cook.* New York: Abelard-Schuman.

Terrae Incognitae, the Journal for the History of Discoveries. 1969-89. Vols. 1-20, continuing. Amsterdam (1969-88) and Detroit (1989-).

Thrower, Norman J.W. 1972. *Maps and Man. An Examination of Cartography in Relation to Culture and Civilization.* Englewood Cliffs, N.J.: Prentice-Hall.

———. 1976. New Geographical Horizon: Maps. In *First Images of America. The Impact of the New World on the Old.* ed. Fredi Chiappelli. Berkeley: University of California Press.

———, ed. 1979. *From Drake to Cook: Two Centuries of British Discovery in the Pacific.* San Francisco: California Academy of Sciences.

———, ed. 1984. *Sir Francis Drake and the Famous Voyage, 1577-1580.* Berkeley: University of California Press.

Tooley, R.V. 1952. *Maps and Map-Makers.* New York: Crown.

———, ed. 1963-75. *Map Collectors' Circle.* Nos. 1-110. London.

———, comp. 1979. *Tooley's Dictionary of Mapmakers.* Preface by Helen Wallis. New York: Liss.

———. 1985. Supplement to *Tooley's Dictionary of Mapmakers.* New York: Liss.

Tooley, R.V., C. Bricker, and Gerald R. Crone. 1968. *Landmarks of Mapmaking.* Amsterdam: Elsevier.

True, David O. 1954. Some early maps relating to Florida. *Imago Mundi.* 11:73-84.

Vietor, Alexander, O. 1962. A Pre-Columbian Map of the World, circa 1489. *Yale University Library Gazette.* 37 (no. 1, July): 8-12.

Vignaud, Henry. 1902. *Toscanelli and Columbus. The Letter and Chart of Toscanelli.* London: Sands.

Von Wieser, R. 1893. Die Karte des Bartolomeo Colombo über die vierte Reise des Admirals. In *Mitteilungen. Institut für Österreichisches Geschichtsforschung, Ergänzungsband* 4. Innsbruck.

Wagner, Henry R. 1931. The Manuscript Atlases of Battista Agnese. *The Papers of the Bibliographical Society of America.* 25:1-110.

———. 1936. *Sir Francis Drake's Voyage Around the World, its Aims and Achievements.* San Francisco: John Howell.

———. 1937. *The Cartography of the Northwest Coast of America to the Year 1800.* 2 vols. Berkeley: University of California Press.

———. 1947. Additions to the Manuscript Atlases of Battista Agnese. *Imago Mundi.* 4:28-30.

———. 1949. Marco Polo's Narrative becomes Propaganda to inspire Colon. *Imago Mundi.* 6:3-13.

Waldman, Milton. 1925. *Americana—The Literature of American History.* New York: Holt.

Wallis, Helen. 1974. Edward Wright and the 1599 World Map. In *The Hakluyt Handbook,* ed. David B. Quinn. 2 vols. The Hakluyt Society, 2nd ser., no. 144. London.

Waters, David W. 1955. *The True and Perfecte Newes of ... Syr Frauncis Drake ... at Sancto Domingo and Carthagena ... 1587 by Thomas Greepe.* Americanum Nauticum 3. Hartford: Henry C. Taylor.

———. 1958. *The Art of Navigation in England in Elizabethan and Early Stuart Times.* New Haven: Yale University Press.

Watts, Pauline Moffitt. 1985. Prophecy and Discovery: On the Spiritual Origins of Christopher Columbus's Enterprise of the Indies. *American Historical Review.* 90:73-102.

Wilford, John Noble. 1981. *The Mapmakers.* New York: Knopf.

Williamson, James A. 1962. *The Cabot Voyages and Bristol Discovery under Henry VII.* With the cartography of the voyages by R.A. Skelton. The Hakluyt Society, ser. 2, vol. 120. Cambridge: The University Press.

Winship, George Parker. 1900. *Cabot Bibliography.* New York: Dodd Mead.

Winsor, Justin. 1889. *Narrative and Critical History of America.* 8 vols. New York: Houghton Mifflin.

———. 1892. *Christopher Columbus. And How He Received and Imparted the Spirit of Discovery.* Boston and New York: Houghton Mifflin.

Winter, Heinrich. 1961. New Light on the Behaim problem. *Congresso Internacional de Historia dos Descobrimentos, Actas.* 2:399-410. Lisbon.

Woodward, David, ed. 1975. *Five Centuries of Map Printing.* Chicago and London: University of Chicago Press.

———, ed. 1987. *Art and Cartography. Six Historical Essays.* Chicago and London: University of Chicago Press.

Wroth, Lawrence C. 1944. The Early Cartography of the Pacific. *The Papers of the Bibliographical Society.* 38: 87-268.

———. 1970. *The Voyages of Giovanni da Verrazzano. 1524-1528.* New Haven and London: Yale University Press.

Yule, Sir Henry. 1926. *The Book of Ser Marco Polo.* 2 vols. New York: Scribners.

Index

Aden, 66
"Admiral's map," 64
Afonso V of Portugal, 2, 12
Africa, 2, 6, 9, 12, 15, 23, 44, 52, 56, 60, 62, 72, 120, 156
Alaminos, Anton de, 76
Albemarle Sound, 88
Albuquerque, 66
Amazon River, 96, 104, 156
America, origin of name, 64
Americae Pars Borealis map (De Jode), 152
Anian, 152
Anticosti, 108
Antigua, 77
Antilia, 9, 23, 98, 152
Appalachian Mountains, 126
Arabia, 56
Arctic, 60
Arctic Ocean, 50, 126
Aristotle, 9
Asia, vi, 7 *passim*
Atlantic Ocean, 1
Atlantis, 9
Avila, 100
Aviles, Pedro Menendez de, 145
Ayllón, Lucas Vasquez de, 40, 72, 84, 96, 100
Azores, vi, 108
Aztecs, 76, 124
Bacalaos, 96
Bacon, Roger, 19
Baffin Island, 156
Bahamas, 26, 120
Balboa, Vasco Núñez, 1, 76, 93, 96, 130
Baltic Sea, 6
Barbosa, Duarte, 77
Bay of Bengal, 66
Bay of Fundy, 120
Beaufort, 84
Behaim, Martin, 15, 18-19
Behaim globe, 39
Beneventanus, Marcus, 50
Bermuda, 61
Bermudez, Juan, 61
Bianco, Andrea, 12
Black Sea, 12, 23, 100
Boazio, Baptista, 140, 144, 145
Bordone, 56
Braun, Georg, 130
braza, 100
Brazil, 34, 60, 72, 84, 96, 98, 116, 120
British Isles, 100, 104
Burma, 67
Cabot, John, viii, 1, 12, 30, 34, 44, 50, 57, 60, 66, 72, 80, 84, 104, 136
Cabot, Sebastian, 77, 84, 92, 104, 136

Cabral, Pedro Alvares, 34, 44, 72
Cadiz, 20, 21
Caicos island, 26
Calicut, 44
California, 124, 126, 132
Canada, 72, 104, 108, 112, 116, 121, 126, 144, 156
Canary Islands, 6, 61
Cantino, Alberto, 1, 34, 52, 64
"Cantino map," 34, 40, 60
Cão, Diogo, 23
Cape Breton, 30, 108
Cape Cod, 102
Cape Fear, 88, 156
Cape Hatteras, 88, 156
Cape Horn, 132, 156
Cape Lookout, 88, 136
Cape Mendocino, California, 156
Cape of Good Hope, 2, 15, 18, 66, 102, 120
Cape St. Vincent, 20
Cape Verde, 120
Cape Verde islands, vi, 23, 34
Caribbean, 140
Carleill, Christopher, 144
Carolina Banks, 98
Cartagena, South America, 1, 124, 140, 144, 145
Cartagena city map (Boazio), 144
Cartier, Jacques, 72, 100, 104, 108, 112, 116, 126
Casenove-Coullon, Guillaume, 20
Catalan Atlas (Cresques), 6-7
Cat Island, 26
Cathay. *See China.*
Cavendish, Thomas, 132, 156
Caveri, Nicolo, 40, 44, 52
Central America, vi, 38-39, 52, 61, 96, 98, 100
Certaine Errors in Navigation (Wright), 126, 156
Ceylon, 44, 66
Champlain, Samuel de, 108, 116, 152
Charles V of France, 6
Charles V of Spain, 76, 77, 81, 92, 104, 124
Chart of the Ocean Sea (Re'is), 62
Chesapeake, 120
Chesapeake Bay, 156
Chile, 130
China, vi, 6-7, 15, 26, 39, 52, 104
China Sea, 67
"Christopher Columbus Chart," 23
Cibola, 152
Cipangu. *See Japan.*
Civitas Orbis Terrarum, 130
Colorado River, 100, 102
Columbia, 120

Columbus, Bartolommeo, 23, 38-39, 144
Columbus, Christopher, vi, viii, 1, 2, 4, 9, 15, 18, 20-21, 23, 26, 28-29, 30, 34, 38, 44, 52, 56-57, 60, 64, 66, 72, 80, 93, 124, 144, 156
Columbus, Diego, 84
Columbus, Ferdinand, 2, 9, 77
Congo River, 23
Contarini, Giovanni Matteo, 44, 50, 51, 56, 60
Coronado, 104
Corte-Real brothers, 1, 44, 66, 93
Corte-Real, Gaspar, 34
Cortes, Hernando, 1, 60, 76, 100, 124
Cosmographie Universelle, 116
Cosmography, 98
Costa Rica, 57
Cresques, Abraham, 2, 6-7
Cresques, Jehuda, 6-7
Cuba, 30, 39, 40, 50, 52, 57, 61, 120, 124
Cuzco, Peru, 1, 121, 124, 130
da Gama, Vasco, 1, 12, 15, 23, 34, 44, 60, 66, 72, 92
d'Ailly, Pierre, 2, 15, 18, 66
d'Anjou, Rene, 20
d'Annebaut, Claude, 112
Darien, 96
Davis, John, 156
Davis Strait, 108, 156
de Coligny, Admiral, 116
Dee, John, 156
De Gourgues, 145
De Jode, Cornelis, 124, 152
De Jode, Gerard, 121, 152
Desceliers, Pierre, 112
d'Este, Ercole, 34
Diaz, Bartholomew, 15, 18, 66
Divers Voyages Touching the Discoverie of America (Hakluyt), 88
Dominica, 61
Drake, Sir Francis, 1, 116, 124, 132, 140, 144, 145, 156
Duke of Lorraine, 52
East Coast of North America, Florida, Greater Antilles map (Le Testu), 116
East Indies, 7, 34, 52, 67, 72, 121, 126, 132
East Indies map (Homem and Reinel), 67
Ecuador, 130
Edisto Island, 84
Edward VI of England, 104
Elcano, Juan Sebastian de, 82
Elizabeth Bonaventure, 140
Elizabeth I of England, 120, 132, 136, 140
England, 9, 44

Eratosthenes, 4
Fagundes, João Alvarez, 108, 120
"Famouse West Indian Voyadge..." map (Boazio), 140
Far East, 4, 6
Ferdinand and Isabella, 20-21, 26, 28, 30, 60
Fernandez, João, 96, 136
Ferrer, Jayme, 7
Finé, Oronce, 96
First Voyage of Columbus, 18-19, 20, 28, 30, 57
Florida, 34, 39, 40, 44, 77, 84, 96, 102, 116, 120, 121, 145
Florida Cape, 108
Fort Caroline, 116
Fourth Voyage of Columbus, 20-21, 38-39, 52, 57, 60
Fra Mauro, viii, 2, 12
François I of France, 66, 88, 108
Frobisher, Martin, 124, 136, 152, 156
Ft. Caroline, 145
Ganges Delta, 66
Garay, Francisco de, 76
Gastaldi, Giacomo, 56, 98, 121
Gemma Frisius, Regnier, 126
Genoa, Italy, 20, 40
Geography (Ptolemy), 50, 64
Gilbert, Sir Humphrey, 136, 156
Gilbert, William, 156
Golden Hind, 132
Gomez, Estéban, 44, 72, 77, 92, 96, 100, 102, 108
Gordillo, Francisco, 84
Granada, Spain, 23
Grand Turk island, 26
Greater Antilles, 120
Greenland, 1, 34, 50, 60, 96, 108, 126, 156
Grenada, Caribbean, 120
Grenville, Richard, 156
Guadaloupe, 61, 62, 77
Guam, 82, 83
Guanahani, 26
Gulf of California (Sea of Cortes), 100, 102, 104
Gulf of Mexico, 40, 76, 96, 100, 124, 140
Gulf of Mexico map (Cortes), 76
Gulf of Siam, 67
Gulf of St. Lawrence, 108, 120, 121
Gymnasium Vosagense, 52
Hakluyt, Richard, 88, 140, 156
Harriot, Thomas, 124, 136, 140, 156
Hawkins, John, 136, 156
Hawkins, Richard, 136, 156
Henry II of France, 112
Henry VII of England, 1, 72, 104

Henry (Navigator) of Portugal, vi, 6, 20
Hipparchus, 4
Hispania, 20
Hispaniola, 30, 52, 57, 61, 64, 124
Hispaniola map (Boazio), 144
Hogenberg, Frans, 130
Homem, Andreas, 120
Homem, Diogo, 120
Homem, Lopo, 66, 67, 77, 120
Hondius, Jodocus, 124, 132, 156
Honduras, 57, 120
Horn of Africa, 66
Huatenay River, 130
Hudson Bay, 104, 156
Hudson River, 156
Hudson Strait, 108
Iberian peninsula, 9, 12
Iceland, 120
Ile D'Orleans, 108
Imago Mundi (d'Ailly), 2, 23
Incas, 130
India, 7, 44, 66, 120
Indian Ocean, 6, 12, 15, 66, 72, 77, 100, 102
India Occidentalis, 60
Indonesia, 67
Introduction to Cosmography (Waldseemüller), 52
Ireland, 9, 120
Isabella, 34
Islamic maps, 12
Isles of the Seven Cities, 23
Jamaica, 30, 38, 52, 124
Jamestown, 156
Japan, vi, 4, 7, 15, 19, 26, 50, 52, 98
Java, 15, 121
Jerusalem, 23
John III of Portugal, 77
Jordan River, 96
Kremer, Gerhard. *See Mercator, Gerardus*.
Labrador, 1, 50, 60, 96, 102, 108, 112, 121, 126, 140, 156
Lacadives, 66
La Cosa, Juan de, 23, 30, 34, 40, 44, 84, 104
Lake St. Peter, 108
Lake Tadouac, 156
Lake Texcoco, 76
Land of Gog, 7
Lane, Ralph, 156
La Roncière, Charles de, 23
La Virginea Pars and *La Virgenia Pars* maps (White), 136
Le Moyne, Jacques, 136, 152
Lesser Antilles, 30, 52, 60, 61, 62
Le Testu, Guillaume, 116
Letter (of Columbus), 28-29

Lima, Peru, 130
Line of Demarcation, 81
Lisbon, Portugal, 20, 34
Lopes, Gregorio, 66
Ludd, Canon Walter, 52
Madagascar, 15
Madeira, 6
Magellan, Ferdinand, 1, 60, 72, 77, 80-83, 92, 98
Magellan's Great Voyage maps (Pigafetta), 80-83
Magellan Strait, 77
Maggiolo, Vesconte de, 60
Magog, 7
Majorca, 6-7, 9
Malacca, 67
Malacca Straits, 67
Malaysian Peninsula, 77
Maldives, 66
Manuel of Portugal, 66
Map of the Discoveries of Columbus (unknown), 28-29
Map of the Equatorial Belt (Columbus and Zorzi), 38
Map of the Indies (Martyr), 60-61
mappaemundi, viii
Mare le Paramatiu, 120
Marine chart and world map (Rosselli), 56-57
Marinus of Tyre, 4
Martellus Germanus, Henricus, 2, 15, 18, 20, 21, 52
Martyr D'Anghiera, Pietro (Peter Martyr), 38, 60-61, 80, 84, 96
Mary of England, 120
Mattan (Mactan) island, 83
Mayaguana island, 26
Mecca, 12, 66
Mediterranean Sea, viii, 4, 12
Mekong Delta, 67
Mela Pomponius, 9
Mercator, Gerardus (Gerhard Kremer), 51, 121, 124, 126, 132
Mercator's projection, 124, 126
Mergui Archipelago, 67
Mexico, 104, 124, 130
Mexico City, 76, 121
Mexico City Plan (Cortes), 1, 76
Miller Atlas, 66
Mississippi River, 76
Moluccas, 77, 81, 102, 132
Molyneux, Emery, 156
Montezuma, 124
Montmorency, Grand Constable de, 112
Montreal, 108
Morales, Andrea, 60
Müller, Johann, 18
Mundus Novus (Vespucci), 50

Münster, Sebastian, 56, 98
Nautical Chart (Pizzigano), 9-10
New England, 102, 126, 144
Newfoundland, 1, 30, 34, 44, 50, 60, 84, 96, 102, 108
New Islands map (Münster), 98
New Mexico, 152
New World map (Ribero and Ramusio), 96
New York, 102
Nicaragua, 38, 57
Nicolaus Germanus, Donnus, vi, 4, 20-21
Niño, 100
Niza, Friar Marcos de, 152
Nombre de Dios, 96, 116
Noronha, Fernando de, 40
North America, vi, 1, 30, 40, 44, 50, 51, 52, 56, 72, 84, 88, 92, 98, 112, 120, 121, 126, 136, 144, 152
North Atlantic, 9, 60
North Atlantic map (Homem), 120
North Carolina, 88
Northern Indian Ocean (Homem and the Reinels), 66
North Pole, 51, 60, 102, 126
Norumbega, 121
Novae Insulae, 98
Nova Scotia, 1, 34, 84, 108, 120, 126
Novus Orbis, 52. See South America.
Nueva Terra de Ayllon, 84
Ocean Sea, 1, 6, 15, 19, 57, 60, 61, 152
Ojeda, 30, 62
Orinoco River, 96
Ortelius (Ortels), Abraham, 56, 121, 124, 126, 132, 152
Oviedo, 96
Ozama River, 144
Pacific Ocean, 1, 82, 100, 120, 124
padrón general charts, 84, 92, 120
padrón real charts, 1, 84
Palos, 20, 21
Pamlico Sound, 88
Panama, 1, 38, 57, 76, 96, 102, 116, 144
Parmentier brothers, 120
Patagonia, 98
Pegu, 67
Peru, 96, 102, 104, 124, 130
Peter of Aragon, 6
Philip II of Spain, 120, 144, 145
Philippines, 82, 83
Phillipps, Sir Thomas, 9, 108
Pigafetta, Antonio, 80-83
Pillars of Hercules, 19
Pineda, Alonso Alvarez de, 72, 76, 96, 100
Pinzon, Vicente, 34

Pizarro, 1, 100, 124, 130
Pizzigano, Zuane, 2, 98
planispheres, 6
Plan of Cuzco map (Braun and Hogenberg), 130
Plato, 9
Plutarch, 9
Plymouth, Massachusetts, 156
Polo, Marco, vi, 2, 6-7, 9, 12, 15, 20, 39, 44, 50, 52, 56, 98, 104
Ponce de León, Juan, 1, 84, 145
portolan sea charts, viii, 5, 9, 23
Port Royal (South Carolina), 116
Portugal, 34, 44
Principal Navigations, Voiages, Traffiques and Discoveries of the English Nation (Hakluyt), 156
Ptolemy, Claudius, vi, 2, 4, 7, 12, 15, 44, 50, 51, 56, 64, 67, 98, 102, 156
Puerto Rico, 52, 124
Quexos, Pedro de, 84
Quinn, David, 136
Quivira, 152
Quiviriae Regnum map (De Jode), 152
Raleigh, Sir Walter, 1, 88, 124, 136, 145, 156
Ramusio, Giovanni Battista, 96, 121, 130
Ramusio map, 96
Re'is, Kemal, 62
Re'is, Piri, 62, 66
Red Sea, 23
Reinel, Jorge, 66, 67
Reinel, Pedro, 66, 67
Ribero, Diego, 1, 77, 92-93, 96, 100, 102, 104, 108
Ringmann, Mathias, 52
Rio de Janeiro Bay, 40
Rio de la Plata, 104, 140, 156
Rio de sa verazanas, 84
Roanoke colony, 1, 88, 124, 136, 140, 145, 156
Roberval, Sieur de, 108
Roberval settlement, 108
Rodadero river, 130
Rosselli, Francesco, 44, 56-57, 102
Russia, 56
Ruysch, Johannes, 50-51, 60
Saguenay River, 108
Sagres, vi
Samana Cay, 2
San Salvador, 2
Santiago, Cape Verde Islands, 140
Santo Domingo, Hispaniola, 1, 92, 124, 140, 144
Sargasso Sea, 9
Sea of Verrazzano, 88, 98, 102, 136, 140, 156
Second Voyage of Columbus, 20, 61
Seneca, 9

Sept-Îles, 108
Seven Cities, 9, 152
Seville, Spain, 20, 21
Sierra Nevada, 121
Smith, John, 136, 152
Solinus, G.J., 9
South America, vi, 38-39, 40, 50, 52, 60, 61, 72, 96, 98, 100, 104, 126, 132
South Carolina, 88
Southeast Asia, 67
South Sea, 76, 98
Spain, 34, 44
Spice Islands, 67, 120, 132
St. Augustine, Florida, 116, 124, 140, 144, 145
St. Augustine map (Boazio), 1, 145
St. Brendan, 9, 62
St. Helena, 84
St. John's River, 116, 145
St. Lawrence River, 104, 108, 112, 121, 126, 140, 156
St. Lawrence Valley, 120
Strait of Anian, 121
Strait of Gibraltar, 120
Strait of Magellan, 96, 102, 112, 126, 132, 140
Strait of Malacca, 67, 77
Strait of Patagonia (Magellan), 82-83
Suleiman the Magnificent, 62
Sumario map, Ribero-Ramusio, 96
Sumatra, 15, 66, 67, 77, 121
Tabula (Ruysch), 51
Talleyrand, 108
Tampico, 76
Temixtitan, 76
Teocalli, 76
Terra Nova, 121
Terre Nove (Waldseemüller), 64
Terrestrial Globe (Behaim), 18
Theater of the World (Ortelius), 121
Third Voyage of Columbus, 20
Toscanelli, Paolo, 2, 15, 19, 61
Travels (Polo), 6-7, 12, 20
Treaty of Tordesillas, 34, 77
Trinidad, 52, 60, 61
Tropic of Cancer, 64
Ulloa, Francisco de, 100, 104
"Vallard" Chart, 108
Venezuela, 60
Vermilion Sea, 100
Verrazzano, Gerolamo da, 40, 84, 88, 102, 112, 116, 124, 126
Verrazzano, Giovanni da, 88, 92
Vesconte, Pietro, 12
Vespucci, Amerigo, viii, 1, 30, 34, 40, 44, 52, 60, 62, 64, 66, 72, 77, 80, 84, 93

Vespucci, Juan, 77, 84

Victoria (Magellan's ship), 77, 82, 98

Virginia colony, 124, 136, 144, 152

Virgin Islands, 52, 60, 61

Waldseemüller, Martin, 34, 40, 44, 51, 52, 64, 104

Watling's Island, 26

Western Europe, 9

West Indies, vi, viii, 4, 26, 28, 34, 38, 44, 60, 62, 64, 120

West Indies map (Martyr), 60-61, 96

White, John, 124, 136, 140

World map
    Agnese, 100-101
    Agnese (oval), 102
    Cabot, 104
    Caveri, 40
    Contarini, 44
    Desceliers, 112
    Hondius, 132
    La Cosa, 30
    Maggiolo, 60
    Martellus, 15
    Mauro, 12
    Mercator, 126
    Ortelius, 121
    Ptolemy, 4
    Ribero, 92
    Rosselli, 56-57
    Ruysch, 50-51
    Verrazzano, 88
    Vespucci, 77, 84
    Waldseemüller, 52
    Wright, 156

Wright, Edward, 124, 126, 156

Yucatan, 40, 77, 96, 120

Zanzibar, 15

Zorzi, Alessandro, 38-39